D1693516

Douglas T. Gjerde, Lee Hoang, and David Hornby

RNA Purification and Analysis

Further Reading

J.S. Fritz, D.T. Gjerde

Ion Chromatography

2009
ISBN: 978-3-527-32052-3

G. Carta, A. Jungbauer

Protein Chromatography

Process Development and Scale-Up

2009
ISBN: 978-3-527-31819-3

P. Herdewijn (Ed.)

Modified Nucleosides

in Biochemistry, Biotechnology and Medicine

2008
ISBN: 978-3-527-31820-9

L.W. Miller (Ed.)

Probes and Tags to Study Biomolecular Function

for Proteins, RNA, and Membranes

2008
ISBN: 978-3-527-31566-6

J. von Hagen (Ed.)

Proteomics Sample Preparation

2008
ISBN: 978-3-527-31796-7

R.K. Hartmann, A. Bindereif, A. Schön, E. Westhof (Eds.)

Handbook of RNA Biochemistry

2005
ISBN: 978-3-527-30826-2

Douglas T. Gjerde, Lee Hoang, and David Hornby

RNA Purification and Analysis

Sample Preparation, Extraction, Chromatography

WILEY-VCH Verlag GmbH & Co. KGaA

The Authors

Dr. Douglas T. Gjerde
PhyNexus, Inc.
3670, Charter Park Drive
San José, CA 95136
USA

Dr. Lee Hoang
PhyNexus, Inc.
3670, Charter Park Drive
San José, CA 95136
USA

Dr. David Peter Joseph Hornby
University of Sheffield
Department of Molecular Biology
Firth Court, Western Bank
Sheffield S10 2TN
United Kingdom

All books published by **Wiley-VCH** are carefully produced. Nevertheless, authors, editors, and publisher do not warrant the information contained in these books, including this book, to be free of errors. Readers are advised to keep in mind that statements, data, illustrations, procedural details or other items may inadvertently be inaccurate.

Library of Congress Card No.: applied for

British Library Cataloguing-in-Publication Data
A catalogue record for this book is available from the British Library.

Bibliographic information published by the Deutsche Nationalbibliothek
The Deutsche Nationalbibliothek lists this publication in the Deutsche Nationalbibliografie; detailed bibliographic data are available on the Internet at http://dnb.d-nb.de

© 2009 WILEY-VCH Verlag GmbH & Co. KGaA, Weinheim

All rights reserved (including those of translation into other languages). No part of this book may be reproduced in any form – by photoprinting, microfilm, or any other means – nor transmitted or translated into a machine language without written permission from the publishers. Registered names, trademarks, etc. used in this book, even when not specifically marked as such, are not to be considered unprotected by law.

Printed in the Federal Republic of Germany
Printed on acid-free paper

Cover design   Schulz Grafik-Design, Fußgönheim
Typesetting   SNP Best-set Typesetter Ltd., Hong Kong
Printing   betz-druck GmbH, Darmstadt
Bookbinding   Litges & Dopf Buchbinderei GmbH, Heppenheim

ISBN: 978-3-527-32116-2

Contents

Preface *IX*
Acknowledgments *XI*

1   RNA Extraction, Separation, and Analysis *1*
1.1   The Need to Be Able to Extract, Manipulate, and Analyze RNA *1*
1.2   Using Chemical Tools to Solve the Problem of Analysis of Biological Processes *3*
1.3   The Principle of Chromatography and Solid-Phase Extraction *4*
1.3.1   Principle of Chromatography *4*
1.3.2   Mobile Phase Gradient Controls Elution *5*
1.3.3   Different Types of Column and Eluent Chemistries *6*
1.3.4   The Principle of Solid-Phase Extraction *8*
1.4   RNA Chromatography *10*
1.5   Enzymatic Treatment of RNA and Analysis *13*
1.5.1   Polyacrylamide Gel Electrophoresis *14*
1.5.2   RNA Structure Probing with Ribonuclease Enzymes *14*
1.6   Content and Organization of This Book *15*
         References *16*

2   Biological and Chemical RNA *17*
2.1   Why Classify RNA with Biology and Chemistry? *17*
2.1.1   Chemical Classification of RNA *18*
2.1.2   Biological Classification of RNA *19*
2.2   Prokaryotic Cellular RNA *20*
2.3   Prokaryote Sample Type *23*
2.3.1   *Escherichia coli* *23*
2.3.2   Other Bacteria *24*
2.4   Eukaryotic Cellular RNA *24*
2.5   Eukaryote Sample Type *27*
2.5.1   Yeast *29*
2.5.2   Other Fungi *30*
2.5.3   Simple Multicellular Organism *30*

RNA Purification and Analysis: Sample Preparation, Extraction, Chromatography
Douglas T. Gjerde, Lee Hoang, and David Hornby
Copyright © 2009 WILEY-VCH Verlag GmbH & Co. KGaA, Weinheim
ISBN: 978-3-527-32116-2

2.5.4	Soft Animal 30
2.5.5	Hard Animal 31
2.5.6	Plant 31
2.5.7	Cell Culture 31
2.6	Other Samples 32
2.6.1	Virus 32
2.6.2	Soil and Rock 32
2.7	Synthetic RNA 32
2.7.1	Aptamers 33
2.7.2	SELEX 34
2.7.3	Short Hairpin RNAs 34
	References 34

3	**RNA Separation: Substrates, Functional Groups, Mechanisms, and Control** 37
3.1	Solid-Phase Interaction 37
3.1.1	Adsorption of Sample Compounds and Sample Matrix Compounds 37
3.1.2	Roles of Solid-Phase Substrate and Functional Group 39
3.1.3	Correlation of Interaction Type, Functional Group, and Substrate 40
3.1.4	RNA Structure and Solid Surface Interaction 41
3.2	The Solid-Phase Substrate and Attachment of Functional Groups 43
3.2.1	Polymeric Resin Substrates 44
3.2.2	Porous and Nonporous Polymeric Resins 45
3.2.3	Monolith Polymeric Columns 47
3.2.4	Functionalization of the Polymer 48
3.2.5	Silica–Glass-Based Substrates 50
3.2.6	Functionalization of Silica 51
3.2.7	Agarose and Cellulose Affinity Substrates 53
3.2.8	Dextran and Polyacrylamide Gel Filtration Substrates 53
3.3	Reverse-Phase Ion-Pairing Separation Mechanism 54
3.4	Ion-Exchange Separation Mechanism 57
3.5	Chaotropic Denaturing Interaction Mechanism 61
3.6	Hybridization 62
3.6.1	SELEX 62
3.7	Gel Filtration 63
	References 64

4	**RNA Extraction and Analysis** 67
4.1	Transcription 67
4.1.1	RNA Catalysis 69
4.1.2	RNA–Protein Complex Interactions 69
4.1.3	Pre-mRNA Splicing 71
4.2	Translation 72

4.2.1	Post-Transcriptional Control of Eukaryotic Gene Expression	76
4.3	Gene Regulation	76
4.3.1	RNA Interference Pathway	76
4.3.2	Micro RNAs and Their Role in Gene Regulation	78
4.4	Use of siRNA to Investigate Gene Function	79
	References	79
5	**RNA Chromatography**	**81**
5.1	Development of RNA Chromatography	81
5.2	RNA Chromatography Instrumentation	85
5.2.1	The Column Oven	85
5.2.2	Ultraviolet (UV) and Fluorescence Detection	86
5.2.3	Fragment Collection	86
5.3	RNA Chromatography Conditions	87
5.4	Temperature Modes of RNA Chromatography	88
5.4.1	Nondenaturing Mode	89
5.4.2	Partially Denaturing Mode	89
5.4.3	Fully Denaturing Mode	90
5.5	Comparison of Gel Electrophoresis and Liquid Chromatography	90
5.5.1	Gel Electrophoresis	90
5.5.2	Liquid Chromatography	93
5.6	Analysis of Human Telomerase RNA Under Nondenaturing Conditions	95
	References	98
6	**RNA Chromatography Separation and Analysis**	**101**
6.1	Features of RNA Chromatography	101
6.2	Separation of Double-Stranded and Single-Stranded RNA	102
6.3	Separation of Cellular RNA Species	107
6.4	Separation of Messenger RNA from Total and Ribosomal RNA	108
6.5	Analysis of Transfer RNA	109
6.6	Chromatography and Analysis of Synthetic Oligoribonucleotides	111
6.7	Application of RNA and DNA Chromatography in cDNA Library Synthesis	114
6.8	DNA Chromatography Analyses of RT-PCR and Competitive RT-PCR Products	117
6.9	Alternative Splicing	120
6.10	Differential Messenger RNA Display via DNA Chromatography	121
	References	124
7	**RNA Structure–Function Probing**	**127**
7.1	Definition of the Structure–Function Paradigm	127
7.1.1	Francis Crick, and Predicting the Existence of tRNA	130
7.1.2	The Discovery of Ribozymes	133
7.2	Structure Determination of RNA	135

7.2.1 Primary Structure   136
7.2.2 Secondary Structure   136
7.2.3 Tertiary Structure   137
7.2.4 Quaternary Structure   138
7.3 Footprinting, Model Building, and Functional Investigations   139
7.3.1 Chemical Probing and Cleavage   140
7.3.2 Modification Interference   143
7.3.3 FRET   144
References   144

Appendix 1
Chromatographic Separation Equations and Principles for RNA Separation   147

Appendix 2
HPLC Instrumentation and Operation   159

Appendix 3
RNA Chromatographic System Cleaning and Passivation Treatment   185

Index   191

Preface

Unlocking the role of RNA in biological cellular processes has proved to be more challenging than perhaps many investigators had first believed would be the case. However, it is also becoming apparent that, as the mysteries of RNA function are revealed, greater rewards will be gained than had been imagined. While life is indeed complicated, this book takes an approach to understanding life by examining it as a set of discrete, simple chemical reactions – the control of which generates the complexity. In continuing with this theme of reductionism, RNA is particularly amenable to *in vitro* analysis and, by isolating the RNA and asking questions in a very controlled manner, the details of specifics relating to the catalysis, affinities and mechanisms of RNA can be identified. As a result, major advances should be achieved in our fundamental understanding of biology, and perhaps even greater medical rewards can be gained.

RNA is fragile and complex in both its structure and function, and advanced tools are required for the reliable capture, purification and analysis of its various types. It is not the purpose of this book to provide a series of "cookbook" procedures for RNA's extraction, separation, and analysis; rather, it is intended as a "tool book", in which we describe the chemical principles that can be used as the foundations for the various methods and tools used to investigate RNA. For that reason, some examples are provided to illustrate the various concepts. By understanding these basic chemical concepts, and how to apply them in terms of the way in which RNA can be manipulated, the biologist should be able to better use the routine products and methods that are currently available, or perhaps develop new techniques for coping with any new-found problems associated with RNA.

In this book, we have brought together the concepts of both biology and chemistry to describe what these tools are, and how they work. Some of the tools described, such as spin columns and precipitation procedures, will already be familiar to the biologist, as will the various procedures described for the different types of RNA. Some other tools may be unfamiliar to the reader, however, and consequently the potential of high-performance liquid chromatography is described in detail, as are the basic instrumentation, the separation columns, and the means by which these separations are controlled.

The special instrumentation and columns required for RNA separations have led to this technology being referred to as "RNA Chromatography", with the breadth of the subject being illustrated clearly by the many applications described.

So, we hope that this book will be used by biologists not only as a means of expanding their arsenal of tools for RNA-based investigations, but also to help them use these tools to great effect. With such techniques available it should, in time, be possible to solve – in creative manner – the vast array of problems encountered in RNA-related research.

May 2009

Douglas T. Gjerde
Lee Hoang
David Hornby

Acknowledgments

The authors thank Dr Chris Suh, Applications Scientist at PhyNexus, who helped with this book by editing and providing information for the text. Chris also drew many of the figures and provided the artwork for the book's cover. The authors also held many philosophical discussions with Chris Suh on RNA in general, and how a book of this type should be written.

Dr Mark Dickman, Professor of Chemistry at the Department of Chemical & Process Engineering at the University of Sheffield, provided valuable information for the book. Mark is an expert in chromatography and mass spectrometry detection, and has very successfully combined these tools with the biology of RNA, as evidenced by the quality of his publications that have been cited in this book, and which he continues to produce. We thank Mark for his valuable information and editorial help.

Mrs Jane Broom, mother-in-law to Gjerde and life-long reader, provided her proof-reading expertise for many of the chapters, and her help is greatly appreciated by all of the authors.

In particular, Lee Hoang wishes to thank Dr Karen Artiles for discussions and thoughtful suggestions on the manuscript, as well as all colleagues and mentors at the Center for the Molecular Biology of RNA at the University of California, Santa Cruz.

Finally, the authors wish to thank their respective families. Each of us has been blessed with families that read and love books. Our families are the most important part of our lives; they provide us with our inspiration and the means to enjoy life. We can only hope that we return the favor.

1
RNA Extraction, Separation, and Analysis

1.1
The Need to Be Able to Extract, Manipulate, and Analyze RNA

RNA is a powerful biological material. It is an active molecule that is directly involved in performing or controlling many biological functions. Unlocking the role of the various types of RNA present in biological cellular processes is critical to developing new methods of diagnosis and understanding and treating disease. New research is being reported daily that someday will lead to new drugs or methods of treating disease. Indeed, some of the work is so promising that it is hoped that a range of diseases will someday be effectively treated and cured. In any case, understanding RNA and its function is one of the keys to understanding life.

There are many different types of RNA, each of which carries out diverse and important functions in the cell, as shown on the book cover and in Figure 1.1. The need to extract and understand the roles of these RNAs is important not only for an understanding of biology but also for developing therapeutics. This particular figure was chosen as the book cover because it illustrates the complexity and importance of RNA in functional mechanisms within the cell. To name a few, RNAs have been shown to function in transcription (pri-RNA, mRNA), splicing (snRNA), translation (tRNA, tmRNA), and post-transcription regulation (siRNA, miRNA). Each type and function of RNA is described in Chapter 2 of this book, while each pathway is discussed in more detail in Chapter 7.

Unlike DNA–which is quite rugged–RNA is transient. RNA is also fragile, elusive, and difficult to recover. Proteins are expressed from DNA using RNA, with the cell carefully controlling which proteins are active, largely through overseeing the synthesis and degradation of RNA. The expression of DNA through RNA to produce new proteins could not be accomplished if the "old" RNA were still present from previous expressions of proteins; consequently, the cells routinely remove the "old" RNA by degrading it with RNase enzymes. Unfortunately, the RNase that is present for biological cellular function can also make RNA collection difficult, and any procedure employed for such collection and use must take this into account.

RNA Purification and Analysis: Sample Preparation, Extraction, Chromatography
Douglas T. Gjerde, Lee Hoang, and David Hornby
Copyright © 2009 WILEY-VCH Verlag GmbH & Co. KGaA, Weinheim
ISBN: 978-3-527-32116-2

Figure 1.1 The schematic arrangement of different types of RNA, and their functions. These include telomerase RNA (teloRNA), primary RNA (pri-RNA), messenger RNA (mRNA), guide RNA (gRNA), small nucleolar RNA (snoRNA), small nuclear RNA (snRNA), ribosomal RNA (rRNA), transfer RNA (tRNA), transfer messenger RNA (tmRNA), signal recognition particle RNA (srpRNA), micro RNA (miRNA), short interfering RNA (siRNA), vault RNA, and catalytic RNA (ribozyme). The arrows indicated alongside each type of RNA show the cellular functions or processes that they are involved in. Each type and function of RNA is described in Chapter 2, and each pathway is described in more detail in Chapter 7.

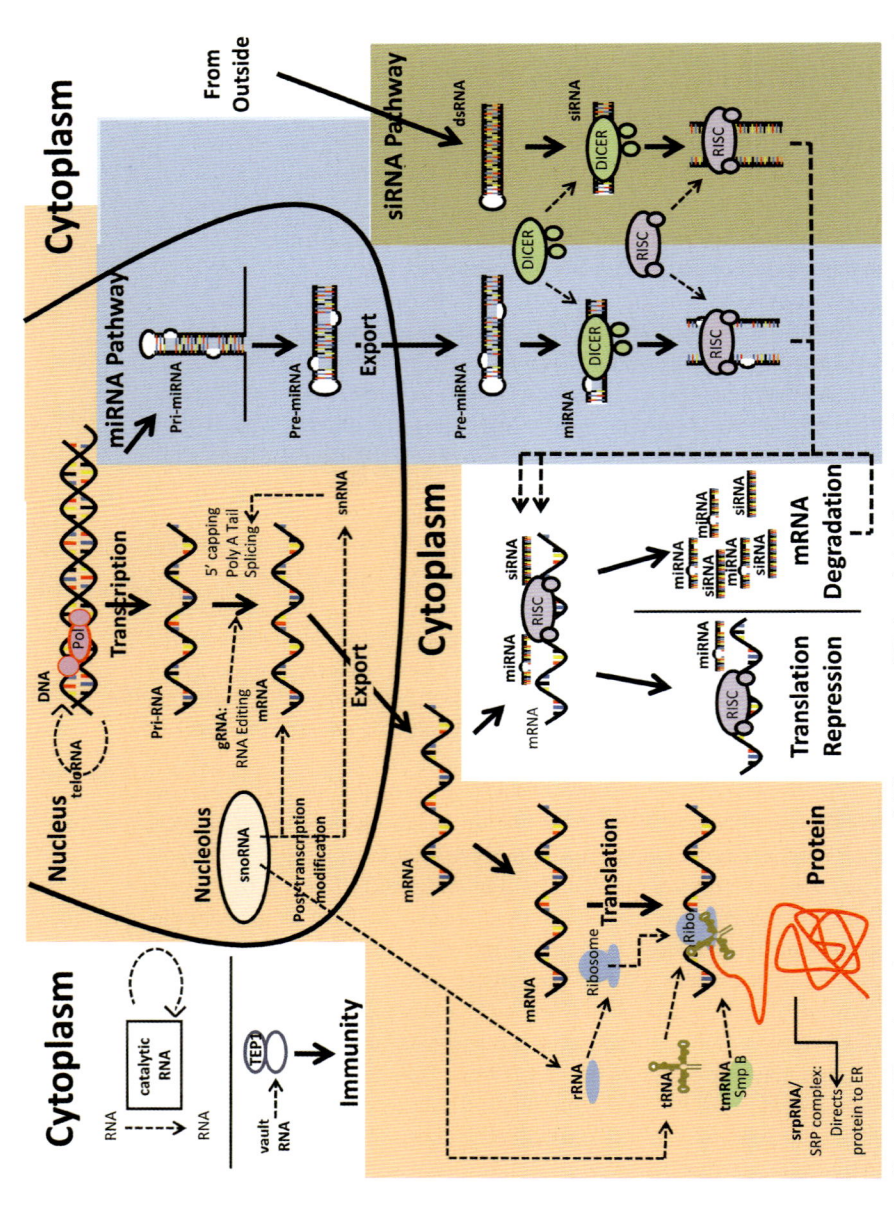

Although many of these procedures have been described, it is not the purpose of this book to provide "cookbook" methods for RNA extraction, separation, and analysis. Yes, there is a multitude of kits available for RNA, and there is also a place for cookbook procedures, as these can free the researcher to use kits for what is tedious and automatic and allow them to design experiments that are of real interest. The danger lies in when a kit does not exactly fit the research problem at hand, or may not even be available. While it may be possible to modify an existing kit to do the job, that will not be possible if the investigator does not understand the chemistry employed in the kit, or how the chemistry can be controlled, manipulated, and optimized to fit the situation.

In this book, we describe the chemistry that is used in the various methods and tools. By understanding the basic chemical concepts, and how they apply to the way in which RNA is manipulated, the biologist will be able to better use the routine products and methods that are currently available on the market. The purpose of this book is also to provide a working knowledge in the context of the types of samples, types of RNA, what is important, and how to accomplish the various tasks through a basic chemical knowledge of the separation and analysis systems. With this information, it is possible for the biologist to make adjustments to fine-tune a product or procedure to a particular research need, or to develop entirely new procedures and methods to fit the problem at hand.

1.2
Using Chemical Tools to Solve the Problem of Analysis of Biological Processes

The types of RNA chemicals tools described in this book are concerned with: (i) solid-phase extraction; (ii) liquid chromatography; and/or (iii) chemical reagents, including enzymes. Both, RNA solid-phase extraction and RNA separation by liquid chromatography operate on similar principles – that is, an interaction of the RNA molecules with the solid-phase surfaces, or within functional groups in the extraction or chromatography media. The surface chemistry of the media can be quite similar – or even exactly the same – in the two cases. Both technologies operate on similar methods of chemical interaction of RNA with the surface. Moreover, both of the methods used to control this interaction and collect working amounts of RNA material are similar.

Part of the difference between solid-phase extraction and liquid chromatography relates to whether classes of molecules are separated, or whether higher-resolution separations are performed on individual molecules. Whereas, solid-phase extraction can be used to extract a particular type or class of RNA, chromatography can be used to separate classes or individual molecules. Nevertheless, both technologies operate on similar methods of chemical interaction of RNA with the surface. The methods used to control this interaction and to collect working amounts of RNA material are also similar.

1.3
The Principle of Chromatography and Solid-Phase Extraction

1.3.1
Principle of Chromatography

The Russian botanist Mikhail Tswett coined the term "chromatography" in 1903, the name being derived from the Greek words "chroma" and "graphein", meaning "color" and "writing" [1]. Tswett described a method for the separation of pigments found in plant leaves by using an open tubular column filled with a dry solid adsorbent of granular calcium carbonate (chalk). To the top of the chalk column, he added an extract of plant material containing its pigments. He then washed the chalk with an organic solvent, which began to flow down through the column by gravity. As the solvent was added to the top of the column, it carried the mixture of plant pigments down the column. Then, as the washing continued, the various pigments began to separate into a series of discrete colored bands, each of which was found to contain a single pigment from the original mixture. Tswett continued to add solvent, so that gradually the regions between the bands became entirely free of pigments and, after a while, all of the bands had been resolved or separated. Tswett then stopped adding the solvent, waited for the solvent to stop dripping, and finally pushed the moist chalk material out of the tube as an intact cylinder. Each pigment could be recovered by cutting the bands apart with a knife and extracting each individual band with an appropriate solvent.

Tswett called this process "chromatography", and was the first to recognize that the separation process was based on a general principle. This was the relative degree by which compounds in solution were either adsorbed to a stationary solid phase of the column or dissolved in the liquid mobile phase. Depending on the liquid phase, a particular compound can partly adsorb to the solid stationary phase and partly dissolve in the liquid mobile phase. Since each compound's affinity for the column is likely to be different, the compound will be carried through the column by the solvent, and the rate that a compound moves down the column will depend on how tightly the compound sticks to the column, or how well it is dissolved in the mobile phase. As the mathematical equations developed later in this book will show, the rate of travel of any one compound is proportional to the ratio of the amount of compound in solution divided by the amount of compound adsorbed on the column, for any given solvent.

This elution process of washing the compounds down the column continues until all components of the mixture are separated (Figure 1.2). Thus, the compounds which compose a mixture can be separated based on how each compound differs in its attraction to the stationary phase or the mobile phase. When a mixture of compounds is applied to the top of a column and washed down the column, there are some compounds in the mixture that do not "like" the mobile phase but rather "like" the stationary phase and so stick tightly or adsorb to the column material. These compounds either stay on the column indefinitely, or are eluted very slowly from the column. Other compounds that "like" the mobile phase and

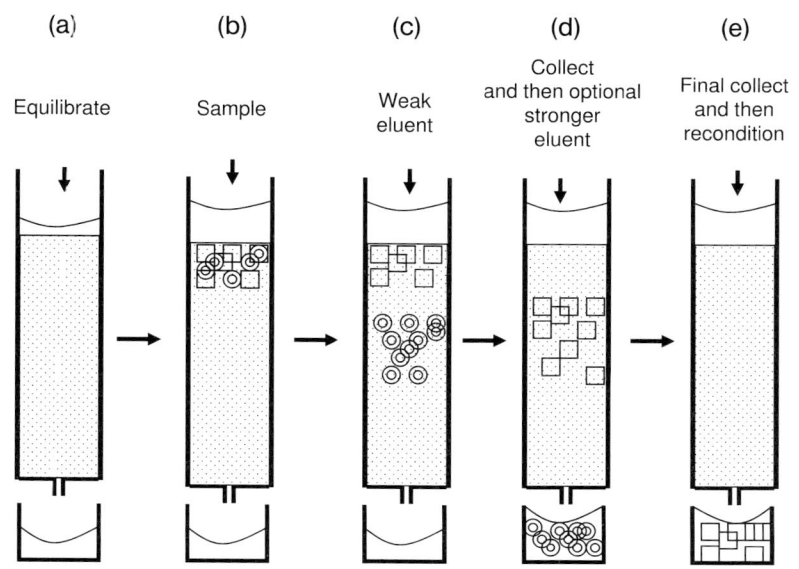

Figure 1.2 The liquid chromatography separation process. (a) The first step is to condition or equilibrate the column to the desired start pH, buffer and/or solvent conditions which ensure that the components are adsorbed into the column. (b) The components of a mixture are added as a concentrated aliquot to the top of the column. (c) Adding eluent to the top of the column starts the separation process. Some components will travel through the column faster because they are solvated by the eluent and interact less with the column packing. Other components interact more strongly with the packing and travel more slowly through the column. Eluent is added from time to time so that the column is not allowed to become dry (the solvent level should not be allowed to fall below the top of the column bed. (d) Component 1 (circles) is eluted and collected. (e) Component 2 (squares) is eluted and collected.

do not adsorb or stick tightly to the stationary phase will travel down the column and be eluted more quickly.

1.3.2
Mobile Phase Gradient Controls Elution

It is quite easy to adjust and control the rate of travel of a compound through a column. The RNA that elutes from a column can be detected by ultraviolet absorbance and recorded either on a chart recorder or electronically. The assessment of the separations involves an analysis of the peaks. Changing the mobile phase to a solvent that does not dissolve the compound as well as the previous solvent will slow the rate of travel of the compounds through the column and improve the separation, albeit at the expense of taking longer to perform the separation. Conversely, a stronger mobile phase will speed up the movement, but (perhaps) at the expense of some of the peaks running down the column together and coeluting.

It is possible to have the best of both situations, however, by using a gradient of eluent strength. When starting with weak initial eluent conditions, most of the compounds will be adsorbed at the top the column and remain there, even with continued washing by the mobile phase. Only those compounds that stick weakly to the column will move down the column and be separated. However, if the eluent strength is then slowly increased it will cause more compounds to move down the column. Such an increase in mobile phase strength is accomplished by changing the solvent to one that can either dissolve or attract the compounds more effectively.

Increasing the mobile phase strength can be carried out in a stepwise manner, by adding aliquots of increasingly strong eluents to the column in succession, with each aliquot addition moving the compounds faster and faster down the column. This procedure is known as a "step gradient" process, because each increase in mobile phase strength is performed as a step. It is also possible, by using the appropriate instrumentation, to increase the gradient gradually–a process known as a "continuous gradient".

1.3.3
Different Types of Column and Eluent Chemistries

It was some time after Tswett's original work before the significance of these discoveries were realized. However, as further investigations were carried out [2–6] it was found that liquid chromatography could be used for many different types of application, including the separation of proteins, organic and inorganic ions, small organic molecules, carbohydrates, and amino acids. The separation of large biological molecules such as nucleic acids came later with refinements in column chemistry and instrumentation.

RNA is one of the most difficult materials to separate under chromatographic conditions. One reason for this is that RNA is an extremely large molecule, and the differences between types of RNA are almost insignificant. RNA molecules are also susceptible to contamination, which causes them to precipitate or to be irreversibly adsorbed to the tubing, frits, columns, or other components of a chromatographic system. Finally, biology works against RNA Chromatography, in that RNA is degraded very quickly in nature and is stabilized only with great difficulty. This includes maintaining stability of the sample before separation, of the RNA molecules during the separation process, and of the collected RNA material when the separation has been completed. The problem here is that ribonuclease (RNase) enzymes are ubiquitous, and their control or removal is necessary to prevent RNA destruction.

The adsorbent contained in the column is based on a core substrate onto which the functional groups that provide column selectivity are bonded. The substrate is chosen based on its uniformity, size, chemical and mechanical stability, and on the ease of attaching appropriate functional groups to the particle surface. Solid-phase substrate materials are usually based on silica, polymer or agarose core particles. Due to problems of inertness and pH stability, a polymer substrate is usually used for the column media in RNA Chromatography, while RNA affinity chromatography will mostly use agarose-based substrates.

Depending on the types of compound to be separated and the desired separation mechanism, ion-exchange, reverse-phase or affinity chemical groups may be bonded to these core substrate particles. All of these groups have been used to separate RNA.

The control of this selectivity is achieved by using different types of eluent to wash compounds down the column, including salts, buffers, acids, bases, and different organic solvents. Each type of separation is based on a particular eluent mobile phase being matched with a column stationary phase, so as to create the correct balance of interaction between the two phases. The selectivity is based on whether a sample compound prefers to be adsorbed onto the column, or to be desorbed and dissolved by the eluent. Specific examples of different combinations of columns and eluents for RNA separation are described in the various chapters and sections of this book. Step gradients or linear gradients, which can be used to alter the eluent composition as the separation is proceeding, can provide additional control on whether a compound moves down the column, or not. It is often advantageous first to retain the sample compound on the column and to wash away all of the impurities. Then, by changing the conditions to increase the dissolving power of the eluent (i.e., by increasing the eluting power of the eluent mobile phase), the sample compounds can be desorbed from the stationary phase and moved down the column.

The classical gravity flow column can serve as a powerful means of liquid chromatography separation. For this, the separation material is made into a slurry with buffer and packed into a column under pressure. For a good separation, it is very important that the resin bed is consistent; in addition, care must be taken to maintain a uniform column when adding sample or buffer to the top of the column, so as to not disturb the column packing. The column must first be "conditioned" by allowing the solvent to run freely through the column, under gravity. Care must be taken not to allow the column to run dry, as this will introduce air into or physically disturb the packed bed and cause an inconsistent liquid flow though the column. (If this were to happen, the liquid would flow around the air pockets, such that any packing material within the pocket would not be exposed uniformly to the sample and eluent.)

After having conditioned the column with solvent, the sample is added to the column top; the liquid mobile phase is then added to the column top and allowed to run through and drip out at the bottom. In some columns, the flow rate can be controlled by using a stopcock. Initially, any nonretained materials will be washed through the column with a wash solvent, after which an elution solvent can be added. If the sample components are colored, a visual record can be used to track the separation and to collect the materials. In other cases, the fractions of eluent can be collected from the end of the column and analyzed, normally plotting the sample concentration versus fraction number; in this way, a chromatogram is constructed.

Under a given set of conditions, a particular sample component will always be found in a certain volume fraction of the eluent, the next compound can be found in the next fraction, and so on. For example, compound A is always located in the 3–5 ml eluent fraction, compound B in the 5–7 ml fraction, and compound C in the 7–11 ml fraction. Thus, under predetermined conditions, even noncolored components of mixtures can be separated, collected in a single fraction, and then analyzed using spectroscopy, titration, and so on, to determine the amounts of each component in the original mixture.

Classical gravity flow liquid chromatography can serve as a powerful separation technique, employing very simple tools. On the negative side, however, the technology can be slow and awkward to use and, frankly, it can also be very boring to watch the drops emerge from a column while ensuring that it does not run dry during any part of the process. Thankfully, the many advances in automation described later in this chapter make chromatography much more accessible to the investigator!

1.3.4
The Principle of Solid-Phase Extraction

The classical liquid chromatography technique described by Tswett is related to solid-phase extraction, but with some important differences. Similar to chromatography, the process of solid-phase extraction involves dissolving the sample in a solvent and passing this through a column containing solid-phase material. Importantly, solid-phase extraction does not rely on chromatography partitioning phenomena (partially sticking and unsticking from the column surface) for separation. Rather, the extraction–separation principle is based on what is sometimes called an either "on" or "off" process of separation, and which generally results in a much more crude purification.

The solid-phase extraction process is shown in Figure 1.3. The first step is to condition the column, when the solid-phase surface is prepared to accept interaction with the sample. This may include treatment at a particular pH with a buffer to convert the surface or functional groups on the solid into a particular chemical form. Alternatively, the conditioning may simply involve using a solvent to wet and clean the surface.

In the next step, the sample containing the compound of interest in a sample matrix is applied to the column. The compound is attracted to the surface with specific binding, and is adsorbed to the column. The selectivity for the sample compound is often very high, with conditions being selected such that all of the material is adsorbed while the matrix mostly passes through the column. The conditions are selected so that matrix materials do not adsorb to the surface (although some may adsorb to a slight degree); this is known as "nonspecific binding" because the selectivity is not high for the matrix material. In contrast, other matrix materials may have a fairly strong, more specific bond with the surface, although the adsorption is still much weaker than that of the sample compound of interest. These materials must be removed in the next step, of washing.

1.3 The Principle of Chromatography and Solid-Phase Extraction | 9

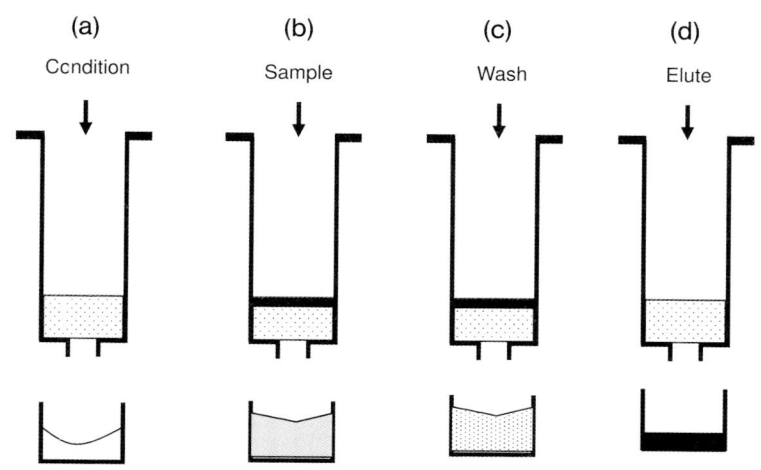

Figure 1.3 The solid-phase extraction process. The column may be a spin column, gravity column, or a vacuum column. (a) The first step is to condition the column. The solid-phase surface is prepared to accept interaction with the sample by equilibration with the appropriate start pH, buffer and/or solvent conditions. (b) In the next step, the sample containing the sample compound of interest in a sample matrix is applied to the column. The sample compound is attracted to the surface with specific binding and adsorbs "on" to the column. The column is usually small, and many of the matrix materials pass through directly through it. (c) The remainder of the matrix materials are removed in the washing step. A strong wash is used if a very pure product is desired, but a weaker wash is used if a high yield is desired. (d) A solvent or buffer is introduced in the elution step, releasing the sample "off" of the column. The purified material is then collected.

If some of the matrix material has a weak affinity for the surface, it may be removed by using a washing solution containing a competitive agent, although of course the wash must be weak enough so as not to remove the sample compound of interest. For the elution step, a solvent is introduced which releases the sample material from the column so that it can be collected in a purified state. In some cases, sample purity may be sacrificed in order to obtain higher yields by using a very weak washing solution; however, if sample purity is preferred, then a stronger washing agent must be used and the yield sacrificed.

Use of the partitioning technique of liquid chromatography allows even compounds with small differences in selectivity for the solid surface to be separated. In addition, by using the on/off technique of solid-phase extraction a particular compound, or class of compounds, can be separated from other classes of material.

Solid-phase extraction is very often performed using so-called "spin columns", which are operated under centrifugal force. Typical spin column bed sizes range from 100 µl to perhaps 0.5 ml, with the smaller columns being the most popular. Spin columns of several milliliter bed sizes are sometimes used when large sample sizes are processed. Normally, flow through the spin column is rapid, and it is necessary to spin the column at relatively slow speeds to direct the sample, wash, and elution solvents through the column. A 5000 r.p.m. spin rate is typical. After each spin the column must be removed and placed in a fresh receptacle before the next solvent is applied to the column. The packing materials are very similar to those used in chromatographic columns. Another approach is to use columns that can either fit onto a vacuum manifold or act as gravity columns. For these types of application, kits are available that include the extraction column and all necessary buffers, prepared in suitable concentrations and volumes.

1.4
RNA Chromatography

Starting in the late 1960s, liquid chromatography changed suddenly and drastically with the development of improved theories and instrumentation. As a result, liquid chromatography was re-named as high-performance liquid chromatography (HPLC). As a process, HPLC became fully commercialized during the late 1960s and early 1970s, when high-performance columns of silica packing materials with bound layers of organic alkyl groups were introduced, along with improved automated pumps and detectors. These new packing materials had a small particle size (average 10 µm), which allowed them to be packed uniformly into small stainless-steel columns. Such columns permitted materials to be separated into very tight bands, which in turn allowed more bands to be separated and smaller elution volumes to be used. In addition, the new instrumentation allowed the use of gradient elution technology to improve the separations.

Liquid chromatographic instrumentation systems employ a pump, an injector, a column, and a detector:

- The pump allows the eluent fluid to be pumped automatically through the column in precise fashion, with uniform and minimal pulsations, accurately, and under high back-pressure conditions (normally 1000–2000 p.s.i.).

- The injector allows the precise and accurate injection of samples into the eluent fluid stream entering the column.

- The detector allows the continuous detection and automatic recording of the separated material eluting from the column.

While developments in the instrumentation are largely responsible for the success of HPLC, the key to the system functioning well is the column, in combination with the interactions of the mixture components with the column and the eluent. To date, most major breakthroughs in chromatographic research have been achieved with the column. Today, the trend is towards even smaller particles, with separation column packings being typically 2 µm in size. In fact, the newest columns contain no particles at all, but are composed simply of a rod of continuous porous polymer monolith filling a small capillary column.

1.4 RNA Chromatography

Both, DNA Chromatography and RNA Chromatography were first described by Guenther Bonn, Christian Huber, and Peter Oefner in 1993 [7–9]. By using ion-pairing, reverse-phase chromatography, this group obtained rapid, high-resolution separations of both double-stranded and single-stranded DNA. Moreover, the separations were usually performed in less than 10 min and, in many cases, a resolution of fragments that differed by only a single base pair in length was achieved. This form of HPLC analysis is largely (though not entirely) based upon the unique separation properties of a nonporous, polystyrene–divinylbenzene polymer bead that has been functionalized with C18 alkyl groups. In this process, an alkylammonium salt (usually triethyl ammonium acetate; TEAA), is added to the eluent to form neutral ion pairs when a DNA sample is introduced into the HPLC instrument. The DNA and RNA fragments are first adsorbed onto the column, and then separated with a gradient of water and acetonitrile, with smaller fragments being eluted from the column first, followed by the larger fragments.

The same technology was also shown to be effective for the separation of single-stranded and double-stranded DNA and RNA. Depending on the type of eluent used, the single-stranded separations are based on differences in the size, polarity and shape of the molecule. By changing the ion-pairing reagent to be more nonpolar, the separation can become mostly sized-based. Temperature is an important parameter for single-stranded separations, especially for RNA, while double-stranded DNA separations are mostly size-based. Double-stranded RNA separations are based on the size and shape of the RNA molecule. A commercially available RNA chromatograph is shown in Figure 1.4.

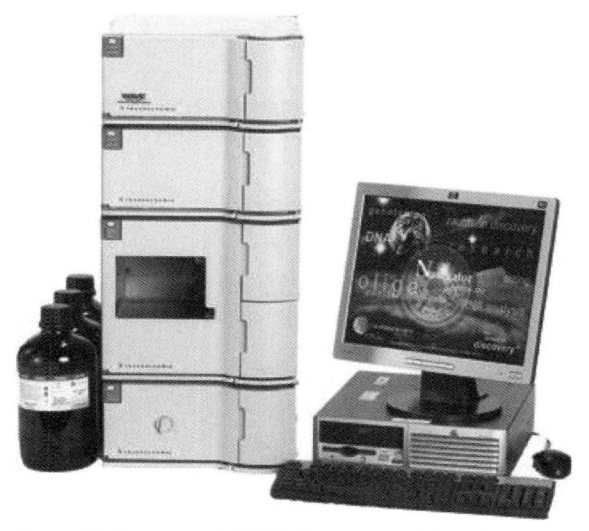

Figure 1.4 A commercial RNA chromatograph based on a high-performance liquid chromatogram designed for the separation and collection of RNA. (Illustration courtesy of Transgenomic, Inc.).

In general, the ion-pairing reagent used is an amine cation salt which forms a nonpolar ion pair with the phosphate anion group of the nucleic acid. TEAA – the most frequently used amine cation salt – is able to pair with the nucleic acid fragments to form nonpolar ion pairs that subsequently adsorb to the neutral nonpolar surface of the column. Acetonitrile is then gradually added to the eluent so as to decrease its polarity until the TEAA/nucleic acid ion-pair fragments are desorbed from the column. In gradient elution, the concentration of acetonitrile pumped through the column is increased gradually as the separation proceeds; under these conditions, the smaller fragments will elute first, followed by fragments of increasing size in line with the increase in acetonitrile concentration.

RNA nucleic acid fragments are extremely large molecules relative to what can normally be separated by a chromatographic process. For example, the average molecular weight of one base pair of RNA is approximately $660\,g\,mol^{-1}$, while a double-stranded, 100 base-pair RNA molecule has an approximate molecular weight of 66 kDa. Typically, 1000 base pair fragments, and up to 2000 base fragments, can be separated using RNA Chromatography.

Although, in this process it is possible to use a very rapid gradient program (e.g., 3–5 min), such a rapid elution program will be possible only at the expense of a lower resolution of the peaks. However, in many cases this will be adequate as the mixture to be separated is not complex. A slower gradient process, of 20–30 min, would produce the highest resolving conditions and the greatest separation of peaks, but clearly under the penalty of a longer analysis time.

One of the most powerful features of RNA Chromatography is that material can easily be purified by collecting it directly from the detector effluent. Notably, the purification of biological samples that might include several types of a particular nucleic acid would be desirable; an example of the detection and subsequent purification of different RNA types from a cell extract is described later in this chapter.

While it is possible to perform such collections by hand, they would normally be accomplished using an automated fragment collector and controlling software, with samples collected into either single vials or large, 96-well plates. In this respect, it is important to account for the dead volume in the tubing which connects the detector to the fragment collector, so as to ensure that the desired peak has actually been collected. The measurement of recovery is performed by taking a small portion of the recovered peak, reinjecting and measuring the area, multiplying this by the ratio of total collected volume to reinjected volume, and comparing this value to the area of the original peak. A typical recovery would be about 80% of the injected material. An additional point is that materials such as RNA may be degraded (via an enzymatic process) after their collection; higher recoveries may not be possible due to a loss of material during the concentration process, whether through precipitation or the plating of material on surfaces.

One very important parameter in fragment collection is the careful execution of the collection times as, invariably, there will be a lag between the time at which the peak is detected and when the fragment reaches the collector probe. Although a timed collection is the most reliable method, unless the timing is correct it is easy to miss some—or even all—of the peak. It is also important to appreciate that there is a dead volume in the tubing from the detector cell outlet to the tip of the deposition probe, and that if this volume is too large it might destroy the resolution of the separation and even result in cross-contamination of the peak of interest with neighboring peaks. It is also important that the probe is cleaned between the collection of peaks, although this is normally carried out automatically by the fragment collector.

1.5
Enzymatic Treatment of RNA and Analysis

RNA has a clear function in every major biological process, including DNA replication, protein translation, and the regulation of gene expression. RNA is an ancient molecule, with much evidence pointing towards it being the first molecule significant for life in the prebiotic world. In recent years, research interest in RNA has been analogous to a chromatographic separation, with periods of great activity followed by some of waning interest. Nonetheless, the new discoveries seem always to "stoke the fire". The first wave of interest in RNA was its role in DNA replication, where short RNA primers—called Okazaki fragments—are required by the enzyme DNA polymerase to copy the DNA template. This was followed by the discovery that tRNA serves as the adapter molecule that forms the bridge between genotype and phenotype. The genetic code embedded in a gene is translated into protein using tRNA, which has specificity for the three-nucleotide codon denoted by the messenger RNA while, at the same time, being recognized by the enzymes that charge them with amino acids. A flurry of interest developed during the late 1980s with the discovery of the catalytic ability of certain RNAs, and this soon led to the discovery of many catalytic RNAs—called ribozymes—and the formation of the RNA World Hypothesis, which states that the first molecules with biologically significant features were RNAs. Most recently, the discovery that RNAs could regulate gene expression at the translational level has unveiled yet another area of biological complexity which might, in time, open the doors to some intriguing and potential concepts of the mechanisms of drug therapy.

With all of the diversity associated with RNAs themselves, and in the processes that they serve, an underlining theme that unites all RNAs is that they have *structure*. In fact, it is the ability of RNA to fold into three-dimensional (3-D) structures that allows it perform such a wide range of functions. Unlike DNA, which usually exists as a double-stranded helix, RNA is normally single stranded. Although its folding mechanism is not completely understood, RNA can form helices, loops, and bulges that provide it with an ability to form long-range intramolecular interactions that give the molecule its shape. Moreover, by understanding the shapes of the various RNA molecules – and how they correlate to function – it should become possible not only to understand many fundamental biological processes but also to provide a basis for rational drug design.

1.5.1
Polyacrylamide Gel Electrophoresis

One of the most powerful methods used in the analysis of RNA is that of gel electrophoresis. In this technique, highly crosslinked and uniform gel matrices are easily generated, with very long gels being capable of separating very small differences between molecules. In fact, the separation ability of polyacrylamide gels allows the resolution of molecules that vary in length only by single nucleotides. In gel electrophoresis, the RNA sample is loaded into the gel at one end, and an electrical current is applied that causes the RNA to move through the gel. For this, the RNA sample mixture is loaded into the gel at one end, and an electrical potential is applied to both ends of the gel. The negatively charged cathode electrode is applied to the sample end of the gel, and the positively charged anode to the far end. When a potential of 1–3 kV is applied, the negatively charged RNA fragments travel through the gel toward the anode. Smaller RNA fragments travel more rapidly through the gel, causing the individual fragment types to be separated on the basis of their size. Polyacrylamide gels have the ability to resolve RNA fragments having only single nucleotide variations in length.

Over the years, a multitude of research groups has used gel electrophoresis to great effect. Typically, a nondenaturing gel can be used to determine if a protein interacts with a particular RNA, for example, in a gel-shift assay. Likewise, RNA splicing events may be analyzed by monitoring (via gel electrophoresis) the extent to which a full-length substrate undergoes splicing to shorten the RNA. Gel separation may also be used, with great sensitivity, for the enzymatic end-labeling of RNA substrates with radioactive phosphate. An alternative choice might be to perform a primer extension reaction, where a primer is annealed to the target RNA, after which the 3′-end and a reverse-transcriptase enzyme are utilized to create a cDNA copy of the RNA. For added sensitivity, radioactive primers or radioactive nucleotide triphosphates can also be used to label the cDNA.

1.5.2
RNA Structure Probing with Ribonuclease Enzymes

As information has accumulated regarding the ability of RNA to fold into 3-D structures, gel separation methods have become very useful. Enzymatic digestion using nonspecific ribonucleases can be used to acquire information as to how an RNA folds and which ligands bind to the RNA by utilizing "footprinting" experiments. Chemical modification agents and primer extension can also be used to determine structural information. As this field matures, new computational tools used to interpret the data tools are being developed.

1.6
Content and Organization of This Book

The tools used to extract, separate and analyze RNA depend not only on the type of RNA being studied but also on its origin. In providing information on this topic, the various types of RNA identified are listed and described in Chapter 2, while the chemical interactions of RNA with solid-phase materials are outlined in Chapter 3. An understanding of these interactions, which include ion pairing, ion exchange, affinity or chaotropic interactions, should in turn allow the manipulation of RNA to be optimized. The various applications of RNA, and methods for its extraction, are described in Chapter 4, while Chapter 5 includes detail of how the RNA Chromatographic process works, and how it compares with classical gel electrophoresis. The various applications of RNA Chromatography are discussed in Chapter 6, together with its general principles and applications, and explanations of *how* and *why* these separations are performed. While many of the details of RNA Chromatography are provided throughout this book, a host of theoretical and practical information is also included in the Appendices.

The different approaches to using chemicals and enzymes in the analysis and interrogation of individual RNAs are described in Chapter 7. One major biological feature of RNA is its ability to fold into 3-D structures, the examination of which is based on: (i) the use of enzymes to cleave the RNA; and (ii) the separation techniques (e.g., gel electrophoresis or chromatography) used to resolve the cleavage products. Today, it is possible to reconstruct the cleavage products and hence to piece together a crude 3-D map of RNA, based on the sites available for, and those protected from, cleavage. This also allows structural information to be acquired when the 3-D structures have not been solved, or to be used in conjunction with the 3-D structures to gain more insight. Further information may also be acquired by studying the cleavage of RNA or RNA–protein complexes, so as to identify the sites of RNA–RNA and RNA–protein interactions.

The Appendices of this book are targeted at those readers wishing to learn more of the theory of liquid chromatography and its practical applications. In Appendix A1, details of the discovery of chromatography and the fundamental concepts that govern its use are outlined. An understanding of these concepts might also help the user to improve separations or to develop new separation conditions. Although mathematical equations describing the fundamental phenomena have been included here, it is important for the new user not to get "bogged down" in these equations if they appear irrelevant; rather, simply read the sections, note the concepts, and then come back when you have gained more experience, and can glean more from the information.

Appendix A2 describes the "nuts and bolts" of HPLC, as well as some details of the various components and functions of the instruments. This information should provide an understanding of how the samples are introduced into, and analyzed by, the instruments. Finally, Appendix A3 outlines the use of HPLC for the separation of nucleic acids. The methodology dictates that the instrument is absolutely free from metal oxide contamination; the reasons for this, and how an instrument can be cleaned and maintained, are listed in this appendix.

References

1 Tswett, M. (1906) Adsorption analysis and chromatographic methods. Application to the chemistry of chlorophyll. *Ber. Deut. Botan. Ges.*, **24**, 384.
2 Poole, C.F. and Poole, S.K. (1991) *Chromatography Today*, Elsevier, New York.
3 Snyder, L.R. and Kirkland, J.J. (1979) *Introduction to Modern Liquid Chromatography*, 2nd edn, John Wiley & Sons, Inc., New York.
4 Heftmann, H. (ed.) (1992) *Chromatography*, 5th edn. *Part A: Fundamentals and Techniques*, Elsevier, Amsterdam.
5 Ettre, L.S. and Meyer, V.R. (2000) Two symposia, when HPLC was young. LC-GC Magazine, July 2000.
6 Horvath, C. (1979) in: *75 Years of Chromatography – A Historical Dialogue* (eds L.S. Ettre and A. Zlatkis), Elsevier, New York, Amsterdam, pp. 704–14.
7 Huber, C.G., Oefner, P.J. and Bonn, G.K. (1995) Rapid and accurate sizing of DNA fragments by ion-pair chromatography on alkylated nonporous poly(styrene-divinylbenzene). *Anal. Chem.*, **67**, 578.
8 Oefner, P.J. and Bonn, G.K. (1994) High-resolution liquid chromatography of nucleic acids. *Am. Lab.*, **26**, C28.
9 Oefner, P.J., Huber, C.G., Umlauft, F., Berti, G.N., Stimpfl, E. and Bonn, G.K. (1994) High-resolution liquid chromatography of fluorescent dye-labeled nucleic acids. *Anal. Biochem.*, **223**, 39.

2
Biological and Chemical RNA

2.1
Why Classify RNA with Biology and Chemistry?

One way to cope with understanding the complexity of life is to take a reductionist's view. The basic building block of an organism – the cell – is composed of a set of four main building blocks, deoxyribonucleic acid (DNA), ribonucleic acid (RNA), the phospholipid bilayer and proteins. These building blocks can be thought of as the core components, which transport genetic information, carry out coordinated enzymatic reactions, and organize and compartmentalize biologically significant molecules including salts, metals, small-molecules and neurotransmitters, and various types of lipids and carbohydrates. To understand the roles of a cell's core components has been the objective of the field of molecular biology for the past 60 years.

There are three approaches to understanding the molecular mechanism of a biological function:

- The *genetic approach* is to mutate the organism and then to scan for a detectable or measurable phenotype. The phenotype is then correlated back to a mutation or mutations in the organism's genome. This method argues that life is complex, and an understanding of a molecular mechanism must be appreciated in the context of all of the complicated interactions that make up a function.

- The *systems biology approach* expands on this idea, and strives to understand the network of interactions that make up life. The interconnectivity of a gene, protein or RNA involves hundreds – if not thousands – of interactions of variable importance.

- In this book, we take a third approach to understanding life by examining it as a set of discrete, simple *chemical reactions*. Understanding these simple chemical reactions, and their control, allows one to build a picture of a molecular mechanism, from the ground up.

The premise for a biochemical study of a biological process is quite simple. First, determine the basic chemical reaction of a biological process. Then, find what comprises the minimal components of this chemical reaction. Purify these components or generate them by an *in vitro* synthesis. Finally, reproduce this chemical reaction using buffers of biological relevance. Besides the physico-chemical measurements that can be monitored – such as kinetic parameters, binding constants and energetic requirements – the basic biochemical reaction may also be subject to interrogation.

The questions to be answered include: What are the biological components that control this reaction? Under what conditions will this reaction take place, and under what conditions will this reaction not take place? In order to build rational conclusions, the components of the experiment must be pure, and all variables must be controlled. To this effect, a study of the biological significance of RNA requires the ability to extract and quantify RNA, and in order to ensure the successful purification and extraction of biologically relevant RNAs, a number of relevant methods will be presented. These methods are based on defining chemical and biological characteristics, which can be used as "molecular handles" with which to capture the RNA.

2.1.1
Chemical Classification of RNA

RNAs have a number of unique characteristics, some of which are suitable for a chromatographic separation while other characteristics pose a challenge for separation. An RNA molecule is simply a polymer. The monomers are made up of a mixture of ribonucleotide monophosphates often called nucleotides for convenience. Ribonucleic acid (RNA) consists of a long chain of nucleotides. Each nucleotide consists of nitrogenous base, adenine (A), cytosine (C), guanine (G), or uracil (U), a ribose sugar with carbon numbered 1' to 5', and a phosphate group attached to the 5' position of the ribose. There are few differences between DNA and RNA. RNA is usually a single-stranded molecule of shorter length compared to DNA, which in addition, is double-stranded. RNA has a hydroxyl group at the 2' position of the ribose while DNA lacks the hydroxyl group. This difference is manifested in the different shape that these molecules adopt when they are base-paired into a double helix. RNA takes on the geometry structure referred as an A-form helix while DNA takes on the B-form. A DNA-RNA heteroduplex will adopt a B-form helix. Another difference between RNA and DNA is that the RNA forms structures that are capable of binding to ligands and/or catalyzing enzymatic reactions while the DNA are carriers of genetic information and require stability in form and function [1].

A number of characteristics allow a researcher to get a molecular handle on the RNA. RNA is polar and electronegative, it is generally larger than protein but smaller than DNA. Like DNA, RNA has a complimentary nature. But because RNAs are usually single stranded, this feature is more accessible. These features make RNA particularly amenable to extraction by ion exchange, ion-paring reverse phase, size exclusion and affinity. A common laboratory practice for separating nucleic acid from protein is through phenol phase extractions and ethanol precipitation under high salt conditions. With variations to these methods, all RNAs can be purified.

2.1.2
Biological Classification of RNA

One approach for purifying RNA uses biological characteristics to define the molecular handles and to develop tools in order to grasp them. Some of the most exciting and current core components of the cell being studied today are the RNA molecules. Although RNA was first recognized as being involved with the flow of genetic information from DNA to protein via messenger RNA (mRNA), an increasing insight has revealed that RNA carries out many enzymatic reactions, for example in the ribosome, and also regulates gene expression at the transcriptional and translational level, using small nuclear RNA and small silencing RNA. To date, investigators have been able to synthesize and evolve RNAs capable of carrying out enzymatic reactions or binding to target ligands with high specificity and high affinity.

The synthetic RNAs of specific function, together with their cellular role in gene regulation, has led to RNA becoming a topic of research and debate not only as a potential therapeutic drug but also as a potential target for therapeutic drugs and diagnostics. In 2004, the Food and Drug Administration approved the first RNA drug, Macugen (developed by Gilead and marketed by OSI Pharmaceuticals and Pfizer) for the treatment of wet, age-related macular degeneration. This single-stranded RNA, called an aptamer, binds and inhibits the function of a protein which is involved with the formation of blood vessels; the inhibition of this protein effectively starves the cells. The unique characteristic that allows RNA to be involved in such diverse cellular roles is its nature to form base pairs, as well as its usually single-stranded state which provides the ability to fold into a three-dimensional (3-D) structure. The 3-D structure of RNA will be discussed further in Chapters 5 to 7.

RNA derives a tremendous amount of diversity from the combination of just four monomers, as seen by their control of gene transcription, by acting at the catalytic core of protein translation, and by controlling gene regulation at the translational level. Whilst this diversity proves to be an efficient way for a cell to reap a multitude of functions from a small set of precursor reagents, there is seemingly little difference between RNA molecules on the chemical level – a fact which poses major challenges for the separation scientist. One major difference between the different types of RNA molecules, such as transfer RNAs (tRNAs) and mRNAs, is that of size. tRNAs are uniform in length at approximately 76–100 nucleotides, while the average length of an mRNA is approximately 1000 nucleotides. Even if it were straightforward, first to separate the total RNA from other cellular components, and second to separate tRNA from mRNA, it would very difficult to separate the individual tRNA species. For these types of separation, the investigator must rely on the unique characteristics of specific tRNAs, such as the anticodon or its specific affinity for enzymes that charge the tRNAs with amino acids, which are known as aminoacyl tRNA synthetases, to perform the separation.

In continuing with the reductionism theme, RNA is particularly amenable to *in vitro* analysis. By isolating the RNA and asking questions in a very controlled manner, specifics regarding catalysis, affinities, and mechanisms can be obtained. The quality of the experiment is in turn based on the quality of the design, and of the reagents, with pure RNA and components being absolute requirements for drawing rational conclusions. One strategy for isolating the particular types of RNA is first to perform a total RNA purification, after which the total RNA fraction is parsed down to the component(s) of interest. In this chapter, we will describe all of the RNAs that exist in these complex mixtures, and outline the possible handles that can be used to separate the components from the mixture.

2.2
Prokaryotic Cellular RNA

The discovery of the structure of DNA marked the beginning of a new age in the field of biology. Prior to the co-mingling of physicists, chemists, mathematicians and geneticists, biology was largely an observational science. Robert Hooke first described the basic unit of life, the cell, by studying and making drawings of cork. The two most famous examples of the power of observation in biology are the laws of heredity and the theory of evolution (seminal studies, to put it mildly!), which were based on acute observations of pea plants and finches, respectively.

Initially, biologists used their observational skills and common sense to classify organisms, with the first biological classification of note dividing organisms into either the animal class or the plant class. The subsequent discovery of single-celled organisms led to a new biological class, the protists, whilst Edouard Chatton later changed the classification based on the presence or absence of a nucleus. These changes in the classification of organisms continued for some time until Carl Woese and others took the decision to classify organisms based on the sequence of their small ribosomal subunit RNAs. This three-domain classification remarkably infers that single-cell yeasts are more closely related to animals than to bacteria. Such a classification system also highlights the fact that RNA is an ancient molecule and that, for the most part, different types of RNAs are conserved from very simple organisms to very complex ones. The evolution of RNA is apparent in the changes in rRNA over time, and makes for a simple correlation to the evolution of a group of related organisms.

There are many types of RNA, with each carrying out different functions in the cell. Generally, most cells contain similar types of RNAs, but differences in the presence or absence of particular RNAs exist depending on cell type, temporal expression, and intracellular localization. The RNAs present in prokaryotes are conserved throughout evolution, with more complicated organisms having added to their RNA repertoire at some point in the evolutionary process and expressing

RNAs not found in the bacteria in order to perform specialized functions. A survey of the RNAs that exist in the prokaryote will set the groundwork for an understanding of the diverse functions of these molecules.

Messenger RNA (mRNA) carries information from DNA to the ribosome, the macromolecular complex where protein synthesis is carried out in the cell. Upon receiving cellular cues, the cell's machinery transcribes genes on the chromosome into mRNA, followed by recruitment to the ribosome for an initiation of translation of the mRNA into protein. This process requires many factors, and is dependent upon GTP hydrolysis. In prokaryotes, the DNA sequence composing a gene is directly copied into an mRNA copy. If the sequence of the gene is known, then affinity separation is possible by immobilizing a complementary RNA or DNA onto a support; this allows for the affinity capture of mRNAs of interest.

Transfer RNA (tRNA) is a small RNA chain of approximately 75–95 nucleotides that is crucially involved in protein translation. The tRNAs deliver specific amino acids to a growing polypeptide chain in the ribosome, as specified by the mRNA codons. The codon is a three-nucleotide sequence that specifies for amino acids, the monomers that constitute the polymeric protein. The tRNA reads the codon by way of its anticodon loop, which is complementary to the codon.

During translation, the tRNA – when precharged with its amino acid – presents its anticodon to the ribosome, which is bound to mRNA; as a consequence, the aminoacylated tRNA is allowed to partially enter the ribosome. The ribosome then positions the tRNA's anticodon in such a way that a cognate codon–anti-codon helix is formed between the codon and the anticodon. If a proper helix is formed, the tRNA is allowed to enter fully the ribosome, where the amino acid it is carrying is used to elongate the growing polypeptide.

There are up to 61 different tRNAs responsible for translating each of the 61 codons, although some tRNAs are capable of recognizing multiple codons, as explained by the "wobble rules". The genetic code also contains three stop codons that are not translated by tRNAs. The basic structure of tRNA is made up of four stems – the acceptor stem, D stem, T pseudouridine C stem, and anticodon stem – and three loops – the D loop, T pseudouridine C loop, and anticodon loop. A number of features are used to distinguish tRNAs. For example, their small size and uniform shape make them possible candidates for purification by size-exclusion chromatography or gel electrophoresis. However, the purification of single species of tRNAs is more challenging, and affinity methods are generally employed. The potential features used for this purification method are the anticodon and the complete tRNA shape, which itself is recognized by the enzymes that charge the tRNAs with amino acids, the aminoacyl tRNA synthetases.

Ribosomal RNA (rRNA) is a major component of the ribosome, which serves as the site for protein synthesis. In prokaryotes, the ribosome is composed up of three RNAs – the so-called 16S, 23S, and 5S (these are named after the Svedberg unit that describes their sedimentation in gradient ultracentrifugation). Together with ribosomal proteins, these rRNAs carry out peptidyl transfer activity and translate the genetic code by using tRNAs and protein translation factors. rRNA is seldom disassociated from ribosomal proteins, and usually exists in a fully folded, globular form. This fact, together with their large size, makes them particularly amenable for purification by ultracentrifugation and chromatography.

Catalytic RNA (ribozymes) are RNAs that catalyze enzymatic reactions. Most of the natural ribozymes that have been discovered break the RNA chains by catalyzing the hydrolysis of phosphodiester bonds. This is often an intramolecular reaction in which the ribozyme self-cleaves. The Hammerhead ribozyme, the Tetrahymena Group I Intron, RNase P RNA, Hairpin ribozyme, HDV ribozyme and VS ribozyme are some of the most common types studied, and have been shown useful in our understanding of RNA folding and the molecular mechanism of RNA catalysis. Ribozymes have a significant role in the formation of the RNA World Hypothesis, which will be discussed in greater detail in Chapter 7.

Vault RNA (vRNA) is a part of a vault ribonucleoprotein complex involved in innate immunity and/or drug resistance. The vRNA–protein complex (vRNP) is composed of three vault proteins, TEP1, VPARP, and MVP, and a small untranslated RNA called vault RNA. TEP1 has been shown to be responsible for stabilizing the vRNAs. Based on structural studies, vRNAs have also been shown to reside at the end of the complex, where the vRNA can interact with both the interior and exterior of the vault particle.

Vault RNA can be extracted using a variation of density gradient centrifugation. First, the cells are lysed, after which a cytoplasmic RNA extraction is conducted. If working with eukaryotic cells, the nuclei must first be removed by centrifugation, and the supernatant is then centrifuged at 100 000 g for 1 h in a sucrose gradient. The intact vault particles (proteins and RNA) are localized to the 40–45% sucrose layer; this layer is first extracted, with phenol extractions being used if the protein component of the particles is required to be removed.

Signal recognition particle RNA (srpRNA) forms part of a signal recognition particle (SRP) complex. SRPs are well-conserved ribonucleoprotein complexes which contain a 300 nucleotide 7S RNA and six proteins (72, 68, 54, 19, 14, and 9 SRPs). SRPs recognize the signal sequence, an amino acid sequence that is read as the peptide is elongating from the ribosome. The SRP binds to this nascent peptide and, upon completing synthesis, the full-length protein is directed to the endoplasmic reticulum in eukaryotes, or to the plasma membrane in prokaryotes.

Transfer-messenger RNA (tmRNA) is RNA that functions as a quality control molecule for translation. tmRNA genes are found in all bacterial genomes. tmRNA stabilizes itself by associating with Small Protein B (SmpB), and has been shown to serve three functions. First, ribosomes stall and cannot complete translation about 13 000 times during the lifetime of a cell. This stalling is sometimes caused by mistakes in the mRNA, which may be too short or contain mutations in which there is no termination codon. The lack of a termination codon means that there is no signal for release factors to bind to the ribosome and break the ribosome–mRNA complex; hence, the ribosomes are bound and stalled. tmRNA is responsible for rescuing the stalled ribosomes by acting as a binding site for these release factors. Second, tmRNA tags the incomplete polypeptide chain that has been stalled during translation. Third, tmRNA promotes the degradation of the aberrant mRNA [2].

2.3
Prokaryote Sample Type

Prokaryotes are a group of organisms that lack a nucleus and other membrane-bound organelles. Most of the prokaryotes are unicellular organisms that reproduce asexually by budding, and are further divided into two classes – Bacteria and Archaea. Bacteria are found in almost all earth environments, including soil, water, and live plants and animals, while Archaea are found mostly in extreme environments such as hot springs and the Earth's deep crust. Prokaryotes also exhibit diverse shapes, and are described by the Latin name for each of these shapes, such as spherical (*cocci*), rod-shaped (*bacilli*), spiral-shaped (*spirilla*), and comma-shaped (*vibrio*).

Although many of the cellular structures are similar between prokaryotes and eukaryotes, there are a few main differences that distinguish the two. As prokaryotes lack true nuclei, their genetic material is not membrane-bound but rather is present in the cytosol in a structure called the nucleoid. Another difference is that prokaryotic DNA exists as a single loop of stable DNA, while eukaryotic DNA is tightly bound and organized into chromosomes.

2.3.1
Escherichia coli

Escherichia coli is a Gram-negative (denoting an absence of an outer membrane) bacterium which is found in the lower intestine of warm-blooded animals. The bacteria are mostly harmless and are approximately 2 µm long and 0.5 µm in diameter. As *E. coli* has a relatively simple genome and is easily grown and manipulated, it has become the best-studied prokaryotic model organism. In modern biotechnology, *E. coli* has played a vital role in development of recombinant DNA technology. By using plasmids and restrictive enzymes, it is possible to produce recombinant proteins using *E. coli*.

E. coli remains one of the most widely studied organisms in the field of biology. Indeed, in many cases the RNA system of *E. coli* is seen to be relevant to higher organisms, because the general mechanism of RNA function has not changed over time. Most of our present understanding of protein translation has derived from studies with *E. coli*, yet the mechanisms are mostly conserved in higher eukaryotes. In those cases where the mechanisms are not conserved, it is necessary to study a different, more relevant organism.

Over time, many RNA extraction methods have been devised, with commercially available kits utilizing the chemical lysis of bacteria. The traditional methods used frequently in the laboratory include sonication and bead beating, although a "French press" may also be used to create a pressure differential that will lyse the cells. Once lysed, the samples are centrifuged to pellet the cell debris, after which RNA extractions are performed using phenol, followed by ethanol precipitation.

2.3.2
Other Bacteria

When *E. coli* studies could not provide the answers to the questions, a variety of other bacteria have been used to study RNA, including the thermophiles – bacteria and Archaebacteria that live in extreme temperatures. These organisms have been studied because they are able to survive at temperatures where the RNAs from *E. coli* would be denatured. Studies involving RNA folding and crystallography both require stable RNAs, for which thermophiles have consequently become an ideal source. When using these different types of bacteria, the RNA extractions may still generally be conducted using the same protocol as for *E. coli*.

2.4
Eukaryotic Cellular RNA

The RNAs that exist in prokaryotes are essentially the same as those in eukaryotes, but can be modified, are sometimes longer, and are regulated by a more complicated machinery. New types of RNAs have also been identified in eukaryotes, these organisms having evolved to carry out more complicated functions, with the RNAs having evolved to follow suit. One reason for the modification of RNAs in higher organisms is the compartmentalization of the cell into the nucleus and the cytoplasm. While the RNAs are synthesized in the nucleus, they must be transported to the cytoplasm for protein synthesis, as is the case for mRNA. In eukaryotes, the mRNAs are tagged with a polyadenylated 3′ end that signals the mRNA for export from the nucleus. Additionally, the 5′ end of the mRNA is capped with a modified guanosine, which serves to regulate transport, degradation, splicing, and translation, while the poly(A) tail serves as an excellent tag for affinity separations of mRNA. rRNAs in eukaryotes are larger than in prokaryotes, and protein translation is eukaryotes is highly regulated at the initiation step, with the mechanisms that ensure fidelity being slightly different from that in prokaryotes.

Telomerase RNA forms part of the telomerase complex, a ribonucleoprotein reverse transcriptase that is responsible for adding specific DNA sequence repeats (TTAGGG) to the 3′ end of DNA strands in the telomere regions of chromosomes. One problem that occurs in eukaryotes, but not in prokaryotes, is that the process of DNA replication involves priming the chromosome with a short RNA that is recognized by the cellular machinery as the site to elongate and copy the chromosomal DNA. As a result, the RNA/DNA hybrid molecule is not stable and the cell will remove the RNA primer. This is not an issue when the RNA primer is in the middle of a chromosome, because the 2′-OH group just upstream of it can be used to prime the DNA polymerase to fill in the sequence previously occupied by the RNA. However, a problem occurs at the ends of the chromosome where there is no available 2′-OH to prime the DNA polymerase, as this causes a shortening of the DNA; this would mean that, for the next round of replication, the chromosome to be copied would be missing its end sequence. The cell's solution to the problem is to generate DNA at the ends of the chromosome, called the telomere. This region is found at the ends of eukaryotic chromosomes which contain condensed DNA material that provides stability to the chromosome. After each replication cycle, the telomeres become shortened; however, telomerase RNA can be used in this region as a template for telomerase to elongate the telomeres.

Telomerase has been implicated in many important topics such as aging, cancer, and other human diseases. The telomerase RNA forms part of the telomerase complex and is often purified using centrifugation methods; however, the sequence of the RNA is known and affinity purification may represent an alternative purification procedure.

Primary mature RNA or **precursor messenger RNA (pre-mRNA)** is an RNA that has been transcribed from DNA prior to being modified to becoming mRNA. Following the Central Dogma of Molecular Biology, mRNA carries information from DNA to the ribosome, where proteins are synthesized. DNA is first transcribed by the enzyme RNA polymerase to form the pre-mRNA, which in turn goes through three main modifications to become a mature mRNA. The three modifications are: (i) 5′ capping for the addition of a modified guanosine; (ii) 3′ polyadenylation for the addition of a poly(A) tail; and (iii) splicing for the removal of introns from the pre-mRNA. During the splicing event, one single pre-mRNA can be alternatively spliced, where different combinations of introns are removed and give rise to many different mRNAs, which in turn gives rise to more diversity of the genome. In eukaryotes, pre-mRNA modifications occur in the nucleus, followed by processing to form the mRNA. The mRNA is transported out of the nucleus to the ribosome in the cytoplasm, where the translation occurs.

In vitro slicing experiments represent a common means of studying splicing. Here, model mRNAs are assayed for a splicing event by polyacrylamide gel electrophoresis (PAGE) when using radiolabeled mRNA substrates, or a proceeding primer extension of unlabeled substrate mRNAs. The substrates used are often model pre-mRNAs that are synthesized by *in vitro* methods using the T7 polymerase to transcribe a DNA plasmid into RNA. The purification of these transcription reactions is commonly performed by phenol phase separations and ethanol precipitation.

Small nucleolar RNAs (snoRNAs) are small RNAs that are associated with RNA maturation in a subcompartment of the nucleus called the nucleolus, and are components of small nucleolar ribonucleoproteins (snoRNPs). snoRNAs are responsible for the post-transcriptional modification of most rRNAs and a few other RNAs, including snRNAs, tRNAs, and mRNAs. They guide methylation and pseudouridylation, with conversion of the nucleoside uridine to a different isomeric form called pseudouridine. These modifications of the different RNAs occur by guiding the snoRNP complex to specific sequences of the target RNA. Each snoRNA acts as a guide for one or two individual modifications in a target RNA. To date, two major types of snoRNA have been classified by a conserved sequence motif: the Box C/D, where C recognizes UGAUGA and D recognizes CUGA; and H/ACA, where H recognizes ANANNA and ACA recognizes ACA.

Small nuclear RNAs (snRNAs) are low-molecular-weight RNAs that are 100–215 nucleotides in length, and are found in the nucleoplasm. They are rich in uridine residues, and are therefore also known as U-RNA. These RNAs are part of a spliceosome complex and, together with protein components, carry out RNA splicing. Five snRNAs – U1, U2, U4, U5, and U6 – have been discovered in the nucleus, while U3 snRNA has been discovered in the nucleolus. These snRNAs do not function as free RNA molecules, but rather form snRNA protein complexes called small nuclear ribonucleic proteins (snRNPs). The spliceosome is a dynamic macromolecule that is composed of different snRNPs, depending upon the function taking place. The spliceosome functions in two major roles that occur with temporal regulation: (i) they are responsible for recognizing the splice site in the pre-mRNA; and (ii) they carry out the transesterification reactions that link exons together to remove intron lariat structures.

The number of spliceosomes in a cell is relatively low, which makes their purification a challenge. *In vitro* analyses of splicing involves working with pure components, such as RNAs that are transcribed or synthesized and recombinant proteins in a reconstitution experiment. Often, spicing studies involve examining what controls the splicing of a model substrate added to nuclear extracts, and avoids any purification of the spliceosome. However, the purification of splicing complexes is especially important for structural studies, when affinity chromatography and other general chromatography methods are can be used to purify the complexes.

Short interfering RNA or **Small interfering RNA (siRNAs)** (also known as small interfering RNAs) are a class of 20–25 base-pair, double-stranded RNAs that are most notably involved in RNA interference (RNAi) pathways, where they interfere with the expression of specific genes. siRNA was first identified in plants as part of post-transcriptional gene silencing (PTGS), and later shown to induce RNAi in mammalian cells [3]. siRNAs are key to the RNAi process where the siRNAs folds into a double-stranded RNA (dsRNA). A dicer enzyme is recruited to this structure and cleaves dsRNA into 20–25 base-pair fragments; one of the two strands is then incorporated into the RNA-induced silencing complex (RISC) and base pairs with target mRNA to form a heteroduplex. The dicer and RISC complex cleaves the heteroduplex, thus regulating the mRNA post-transcriptional gene silencing. The RNAi pathway has been studied extensively, not only to provide an understanding of this fundamental biological process, but also as a possible mechanism for therapeutic drug action. Notably, these siRNAs can be synthesized.

Micro RNA (miRNA) is a single-stranded RNA of 21–23 nucleotides in length, that regulates gene expression. miRNA is transcribed from DNA as a long primary transcript RNA (pri-miRNA) with a cap and a poly-A tail. pri-miRNA is processed in the nucleus by shortening to a 70 nucleotide stem–loop structure called pre-miRNA, followed by further processing to form a mature miRNA in the cytoplasm by interacting with the endonuclease, dicer. Mature miRNAs function to regulate translation by binding to a 3′ untranslated region (UTR) of a target mRNA by a partially complementary fashion, where the miRNA–mRNA duplex is not a complete helix making bulges. This formed duplex structure between miRNA and mRNA can lead to two downstream effects: (i) the formed duplex can prohibit translation without degrading mRNA, thus downregulating gene expression; and/or (ii) the formed duplex may recruit dicer and other proteins to initiate formation of the RISC, which goes through the RNAi pathway where the mRNAs are degraded. Although miRNAs were first discovered using genetics studies, they are now mainly investigated using bioinformatic methods and confirmed experimentally. These short RNAs may be easily synthesized and, in order to study the global population of miRNAs in a tissue or in a group of cells, the miRNAs may be purified from the total RNA by using size-exclusion separation techniques.

Guide RNA (gRNA) is a small RNA that guides the insertion of uridines into mRNAs in a process called "RNA editing". It has been shown that the 5′ end of a gRNA hybridizes to a short region called the "anchor sequence" in pre-mRNA, while the 3′ end functions as a template for the editing process.

Aptamers are RNAs that bind to ligands. They can be synthesized in the laboratory to bind to specific targets, and have evolved to exhibit high affinity and selectivity. Aptamers have great potential as therapeutic drugs, and are discussed in more detail in Section 2.7.1. The first aptamer to be discovered with a biological function was the riboswitch [4]; these are transcribed at the ends of mRNAs and function as small molecules. The presence or absence of a small molecule serves to regulate the translation of that particular mRNA.

Non-coding RNA (ncRNA) is defined as any RNA that is not translated into a protein. They are also known as non-protein-coding genes (npcRNAs), non-messenger RNAs (nmRNAs), or functional RNAs (fRNAs).

Small hairpin RNA (shRNA) is a sequence of RNA in which the helix is joined by a short loop called the "hairpin". shRNA can function to silence gene expression by RNAi. The shRNA is introduced to the cell where the shRNA hairpin structure is cleaved by cellular machinery into siRNA which then functions through RNAi pathway to knock down mRNA to which it has base-paired.

2.5
Eukaryote Sample Type

There are a few fundamental differences between prokaryotes and eukaryotes with regards to RNA (Figures 2.1 and 2.2):

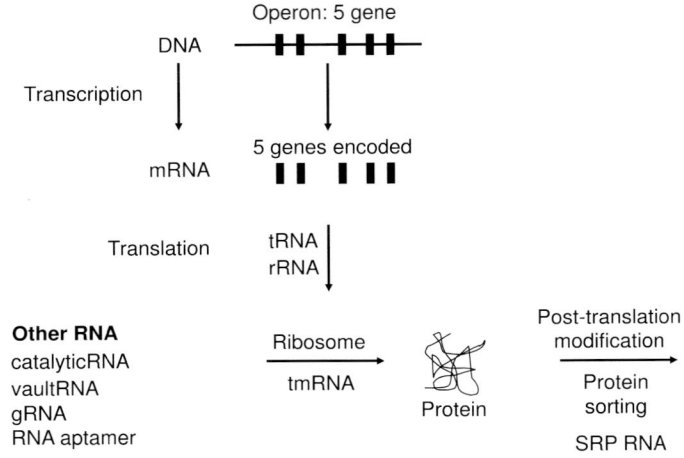

Figure 2.1 Similar RNAs are found in both prokaryotes and eukaryotes. The main differences are that prokaryotic RNAs are mostly maintained throughout evolution, but the eukaryotic RNA repertoire has increased to accommodate more complexity. Prokaryotic RNAs are synthesized during transcription to produce messenger RNA (mRNA). The ribosomal and transfer RNAs are utilized during the translation of mRNA into protein. Transfer-messenger RNA (tmRNA) is used to rescue ribosomes that have stalled during the translation process. The signal recognition particle, an RNA–protein complex, is involved in signaling and the transport of appropriate proteins to the plasma membrane. A number of small RNAs, including catalytic RNA, vault RNA, guide RNA, and aptamers, are involved in important processes in the cell.

- Prokaryotes have less complex RNA types.
- Telomerase RNA, pre-mRNA, siRNA, snRNA, snoRNA, gRNA and miRNA are not present in prokaryotes.
- mRNA, tRNA, rRNA, tmRNA, SRP RNA, catalytic RNA, vault RNA, are present in both prokaryotes and eukaryotes.
- Prokaryotes lack a nucleus; hence, the subcellular localization of the different types of RNAs will differ from that of eukaryotes. For example, mRNA, tRNA, rRNA, and snoRNA may be found inside the nucleus in eukaryotes, whereas in prokaryotes they are found only in the cytoplasm.

Eukaryotes cover a wide range of organisms from different groups, such as animals, plants, fungi, and protists. Among other differences between prokaryotes and eukaryotes, eukaryotes are classified as organisms with a true nucleus, while prokaryotes lack a nucleus; consequently, the process of cell division will differ between the two. Eukaryotes contain a cytoskeleton, composed of microtubules, microfilaments, and intermediate filaments, which plays a vital role in cell division. Eukaryotes are typically larger and contain a variety of internal membranes and structures called organelles, such as the nuclear envelope, endoplasmic reticulum, Golgi body, and lysosomes, but prokaryotes lack these structures.

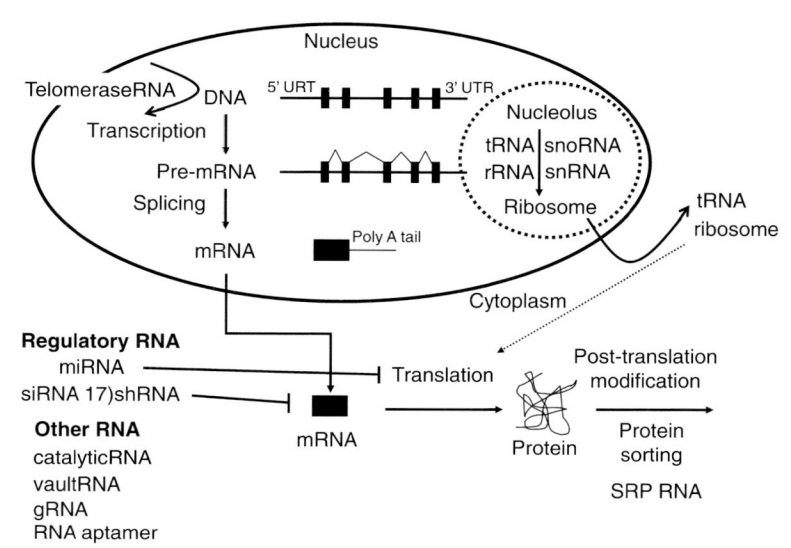

Figure 2.2 Eukaryotes utilize RNAs that are similar to prokaryotic RNAs, but the increased complexity of these organisms results in more RNAs performing diverse functions. Telomerase RNAs maintain chromosome integrity. A whole group of RNAs located in the nucleolus is involved with adding modifications to tRNA, rRNA, and snRNAs. The spliceosome complex is involved in processing pre-messenger RNAs, while small RNAs, the silencing and micro RNAs, are involved with gene regulation at the transcriptional level. These different RNAs highlight the complexity of functions the central role that these molecules play.

2.5.1
Yeast

To date, approximately 1500 different yeast species have been identified and described. Yeasts are unicellular organisms, the majority of which reproduce asexually by budding, although a few reproduce by binary fission. S*accharomyces cerevisiae*, the yeast used in baking and fermentation, has for many years been a model organism in a wide variety of areas in biological research. *S. cerevisiae* requires organic compounds as a source of energy, and can easily be cultivated in the laboratory at 30 °C. *S. cerevisiae* also has a very similar cell cycle to humans, and performs similar basic mechanisms such as DNA replication, recombination, cell division, and metabolism as do human cells. The yeast may be easily manipulated, and are often used in a technique known as the "yeast two-hybrid", which is used to test the interaction between two genes.

Two different methods of cell lysis may be used to isolate RNA from yeast. In the first method, the RNA is isolated directly from intact yeast cells by extraction with hot acidic phenol. This is the most frequently used protocol as it is convenient and little variation is observed from sample to sample. The second method relies on disruption of the cells by vigorous mixing with glass beads and denaturing reagents.

2.5.2
Other Fungi

Although fungi are largely similar to animal cells, they do show some differences. First, fungi have a cell wall material – chitin – which is similar to that in plant cells. Chitin requires special attention during RNA extraction processes, as it is strong and difficult to lyse. Second, higher fungi contain a porous structures called a septum – a passage between two cells that allows the transfer of cytoplasm. Third, some primitive fungi, *chytrids*, utilize a flagellum for their movement. There is a great diversity among the different forms of yeast, with remarkable variation in both cell size and shape. Some yeast are also polynucleate, which means that a large cell may contain many nuclei. Yet, despite these differences, the RNA purification strategies used for yeasts are virtually identical.

2.5.3
Simple Multicellular Organism

The sea sponge, classified as the phylum Porifera, is among the oldest known animal fossils. Sponges form a vital part of coral ecosystems, their health being an excellent indicator of water quality, especially around populated coastal areas [5]. The tissues from sponges can be collected, macerated in a buffer, and the cell homogenate centrifuged to form a pellet, such that the supernatant contains the RNA. The RNA may subsequently be purified and extracted using a phenol–chloroform method.

2.5.4
Soft Animal

Caenorhabditis elegans is a small, free-living, soil nematode which is about 1 mm in length, and was chosen as a model organism during the 1960s by Sydney Brenner to study developmental biology. This small organism has attracted huge research interest because it can be cultivated easily in the laboratory, and there are many techniques and tools available to study its development at the subcellular level. Today, by using *C. elegans* as a model organism, major advances continue to be made in studies of longevity, neurodevelopment, genomics, cell cycle research, and RNAi.

While many protocols have been developed for the extraction of RNA from *C. elegans*, a major problem here lies in efficiently lysing the organism's hard outer layer, the chorion. The most popular lysing procedures involve the use of tissue homogenizers, or even bleach. When the chorion has been lysed, the RNA can be extracted essentially by using the same methods as for other organisms. This includes purifications though cesium gradients, or by using phase separations involving phenol–chloroform in the presence of guanidinium.

2.5.5
Hard Animal

Drosophila melanogaster is a common fruit fly that is perhaps the most commonly used model organism in biology. It is particularly amenable to the study of development by way of genetics and biochemical methods. For RNA studies, *Drosophila* is used to study the RNAi pathway.

The method of extracting RNA from *Drosophila* is largely similar to that for *C. elegans*, although specific tissues may sometimes be dissected prior to extraction so as to study their RNA content. In this process, the tissues (or even whole flies) are flash-frozen and homogenized in buffer, using a mortar and pestle. A phase extraction method is useful here, both for the purification of RNA and the removal of RNases.

2.5.6
Plant

Arabidopsis is a model plant organism used in biological research that, due to its relatively short life cycle, has undoubtedly become one of the most well-characterized. The major advantage of using *Arabidopsis* is the ability to obtain large quantities of RNA, as essentially unlimited amounts of starting material are available. So, if more purified material is required, the simple solution is to use more plants! The protocol for RNA extractions with *Arabidopsis* is based on the methods devised by Anne Krapp and coworkers [6]. For this, 1–2 g of fresh material is freeze-dried, ground to a powder, and then homogenized in buffer (4 M guanidinium thoicyanate, 20 mM EDTA, 20 mM MES, 50 mM 2-mercaptoethanol, pH 7.0). After homogenization, the samples are centrifuged at 8000 r.p.m. (5000 g) at 4 °C for 10 min to pellet the cell debris. The supernatant may then be subjected first to a normal phenol–chloroform extraction to remove the protein components, and the RNA recovered by ethanol precipitation.

In order to extract RNA from plants rich in polyphenols and polysaccharides, the cationic surfactant cetyltrimethylammonium bromide (CTAB) is added to the extraction buffer at a final concentration of 2% [7]. The CTAB will be removed in the subsequent phenol–chloroform extraction and ethanol precipitation steps.

2.5.7
Cell Culture

Today, many different types of cell culture are available for use in biological research, with cultures of insect or mammalian cells providing a wealth of possibilities. The cultured cells should be processed immediately after their removal from the incubator, by centrifuging the cell suspension and discarding the supernatant. When using cells grown in monolayers, the culture medium must be removed prior to processing. To the pelleted cells should be added a denaturing solution [e.g., 4 M guanidinium thiocyanate, 25 mM sodium citrate, pH 7.0, 0.5% (w/v) *N*-laurosylsarcosine (Sarkosyl) and 0.1 M 2-mercaptoethanol], and the lysate then resuspended by agitation. This will cause the DNA to fragment, thus minimizing any contamination by DNA in the phenol extraction.

2.6
Other Samples

The purification of RNA has many uses outside the research laboratory setting. For example, the purification of RNA from a blood sample in a clinical laboratory may require procedures to be used that are both robust and consistent. For this, the technician must follow procedures in a strict manner, with as minimal variation as possible. For these procedures, commercially available kits are the tool of choice, as their ease of use, combined with excellent quality control, will provide procedures that are well-suited to such an environment.

2.6.1
Virus

Unlike prokaryotes and eukaryotes, some viruses contain only genomic RNA and/or catalytic RNA, both of which are employed as the genetic material when taking over host cells in order to reproduce (Figure 2.3). The genomes of these viruses are interesting, with the reverse transcription procedure for producing complementary DNA copies of the virus RNA genome for cloning being especially useful. With viruses, the RNA is extracted from the protein capsid prior to the reverse transcription reaction occurring. Notably, the use of guanidinium and phenol–chloroform is relevant here, to remove any RNases present.

2.6.2
Soil and Rock

The extraction of RNA from soil and rock samples follows the same procedure as for purification from tree samples. The addition of CTAB improves the extraction of RNA from soil and rock; typically, the CTAB concentration in the extraction buffer should be raised to 5%, so as to improve the efficiency of the process.

2.7
Synthetic RNA

RNA is a polymer of nucleotide monophosphates which, in the cell, are polymerized by nucleotide triphosphates being used as substrates. However, for chemical synthesis *in vitro*, several alternative strategies may be employed to ensure both efficiency and accuracy. For example, machines used for oligonucleotide synthesis

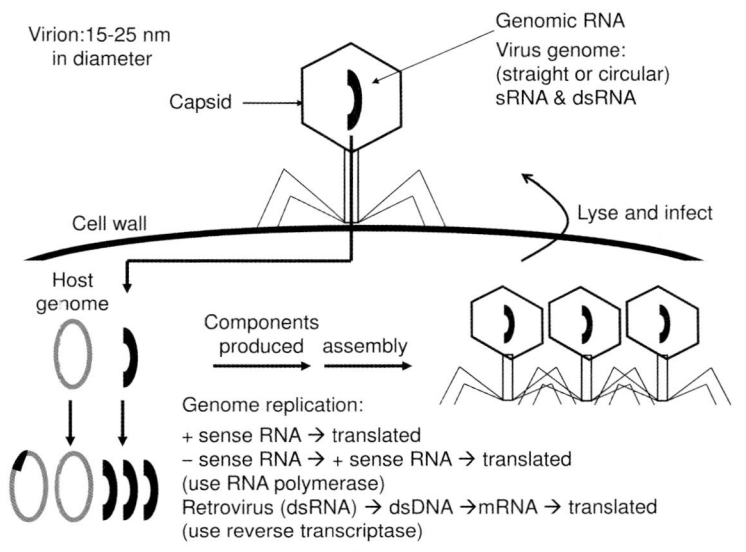

Figure 2.3 Some viruses use RNA as the genetic material. These viruses use the cell's machinery to replicate its genome and make essential proteins that ensure the virus will survive and proliferate. This figure highlights the steps involved in replicating the RNA virus genome. One of the proteins involved in this process is the reverse transcriptase enzyme, which creates a complementary DNA copy of the virus RNA genome (this enzyme is also very useful in the research laboratory setting).

generate RNAs in an automated fashion, where the user simply programs the sequence of RNA to be generated, attaches a solid support column with the 3′ nucleotide immobilized, and then loads on the appropriate reagents. The synthesizer deprotects the 3′ nucleotide, and then introduces a phophoramadite that will covalently bind to the growing polymer via an S_N2 reaction. The unreacted substrate is then washed away, and the cycle continues. Following synthesis, the 2′ hydroxyl groups are deprotected and the immobilized RNA is cleaved from the column.

2.7.1
Aptamers

Aptamers, which are RNAs capable of binding to ligands, were first developed in the laboratory during the 1990s by using the SELEX process (see Section 2.7.2). These short RNAs fold into a 3-D shape that provides them with specificity for target molecules. Single-stranded DNA was later discovered to exhibit an ability to fold into aptamers. The first naturally occurring aptamer – the "riboswitch" – was discovered in 2002, in the laboratory of Ron Breaker. The ease by which an RNA could be evolved with tight binding properties and a high specificity for targets has led to aptamers becoming an intriguing potential therapeutic and diagnostic molecule. The first RNA aptamer to be approved by the FDA as a drug was

Macugen; this was an anti-VEGF inhibitor designed to serve as an anti-angiogenesis drug for the treatment of wet, age-related macular degeneration.

2.7.2
SELEX

The process of a systematic evolution of ligands by exponential enrichment (SELEX) was developed to generate RNA molecules with desired characteristics. For example, if a protein was the desired target, then a pool of RNAs of random sequence would be selected for binding to the immobilized target. The majority of the RNAs could be washed away, while those which exhibited a certain binding characteristic could be recovered. These positive sequences could then be passed through a series of mutageneses, using low-fidelity PCR methods, and subjected to additional rounds of selection. On completion of the evolution process, a sequence would be obtained that not only had the desired function but was also optimized for that function.

2.7.3
Short Hairpin RNAs

Small RNAs, which may be chemically synthesized, are valuable not only for studying RNA folding but also for performing biophysical measurements of stability. The ability of a nucleic acid to form a helical structure and base pairs was first noted from studies with DNA; in the case of RNA, however, the molecule was able to form helices, despite the helical structure being a different shape, due to the 2′ OH group and the conformation of the ribose. The structure of RNA is also interesting in its ability to form 3-D structures. Although the RNA is constrained by bond angles, it can adopt very sharp turns and change direction in as little as four nucleotides. These short hairpin RNAs may be used to study the specific nucleotide sequences required to form these turns, and to identify the rules determining the folds.

References

1 Ab, G., Roskam, W.G., Dijkstra, J., Mulder, J., Willems, M., Van der Ende, A. and Gruber, M. (1976) Estradiol-induced synthesis of vitellogenin. III. The isolation and characterization of vitellogenin messenger RNA from avian liver. *Biochim. Biophys. Acta*, **454**, 67.

2 Richards, J., Sundermeier, T., Svetlanov, A. and Karzai, A.W. (2008) Quality control of bacterial mRNA decoding and decay. *Biochim. Biophys. Acta*, **1779**, 574.

3 Elbashir, S.M., Harborth, J., Lendeckel, W., Yalcin, A., Weber, K. and Tuschl, T. (2001) Duplexes of 21-nucleotide RNAs mediate RNA interference in cultured mammalian cells. *Nature*, **411**, 494.

4 Nahvi, A., Sudarsan, N., Ebert, M.S., Zou, X., Brown, K.L. and Breaker, R.R. (2002) Genetic control by a metabolite binding mRNA. *Chem. Biol.*, **9**, 1043.

5 Donaldson, K.A., Griffin, D.W. and Paul, J.H. (2002) Detection, quantitation and

identification of enteroviruses from surface waters and sponge tissue from the Florida Keys using real-time RT-PCR. *Water Res.*, **36**, 2505.

6 Anne Krapp, B.H., Schäfer, C. and Stitt, M. (1993) Regulation of the expression of *rbc*S and other photosynthetic genes by carbohydrates: a mechanism for the 'sink regulation' of photosynthesis? *Plant J.*, **3**, 817.

7 Chang, S., Puryear, J. and Cairney, J. (1993) A simple and efficient method for isolating RNA from pine trees. *Plant. Mol. Biol. Rep.*, **11**, 113–16.

3
RNA Separation: Substrates, Functional Groups, Mechanisms, and Control

3.1
Solid-Phase Interaction

Extraction and liquid chromatography are based on the interaction of RNA with a (stationary) solid phase that is moderated or controlled by a (mobile) liquid phase passing through the column. As noted in Chapter 1, similar or even identical solid-phase separation media can be used for both solid-phase extraction and liquid chromatography. This solid phase can have many different names, including separation media, separation phase, stationary phase, column packing, column phase, and so on.

The solid phase can take many different chemical forms and be used in different ways to produce the desired separation. Some separations may be enhanced by the attachment of functional groups to the solid-phase substrate or matrix. The choice of the solid-phase substrate, functional group, and the mechanism used for separation, depends on the sample compounds and the desired outcome of the separation, and perhaps also the limitation imposed by the sample in terms of the sample matrix and relative concentrations of components. While there may be no general rule in selecting the solid-phase substrate, functional group and control mechanism for a particular desired purification, it is helpful to understand the different types of solid-phase extraction and their mechanism(s).

3.1.1
Adsorption of Sample Compounds and Sample Matrix Compounds

The solid-phase adsorbent contained in a column is based on a core solid-phase substrate onto which the functional groups have been bonded. In order for the column material to be effective, it must be selective; this means that the column adsorbent must have two important properties: (1) that the adsorbent interacts with and adsorbs the sample components of interest; and (2) that the adsorbent does not interact with or adsorb the sample matrix components that are not of interest.

Although these two adsorbent properties are separate and distinct, both are important. For example, an adsorbent that interacts with some of the compounds of interest, but not all of the sample compounds of interest, may have its uses – but it will also have its limitations which generally are obvious. For example, a sample compound of interest cannot be captured if the solid phase is not selective for that particular compound. However, these limitations are not uncommon and therefore no assumption on column performance should be made unless the manufacturer has specified that the material works in the manner desired. In cases where the adsorption selectivity is limited, it should be understood which sample compounds are absorbed, and under what conditions, before it is considered. Equally, an adsorbent that captures all materials – including the sample compounds of interest and sample matrix compounds – is completely useless. Even if the sample matrix materials could be subsequently and selectively washed from the column, leaving the sample compounds of interest, the column capacity could be easily comprised and overloaded.

A column that only adsorbs major matrix compounds but is not selective for sample compounds is of questionable value, because there can be no certainty that the depletion process will not change the sample. These so-called "depletion columns" are sold by several companies, but no guarantees are made with regards to retaining sample integrity.

There are two possible mechanisms by which depletion columns can harm a sample. In the first mechanism, the depletion of a major matrix component may destabilize the sample so that, once the major component has gone from the sample, the minor components of interest remaining in the sample can precipitate or plate out on a surface. RNA may be particularly susceptible to this effect, due to the phosphate groups that can form insoluble complexes with certain residual metal ions such as iron, copper, and chromium (this subject is discussed in Chapter 5 and Appendix 3). These trace metal ions, if present, may become available for RNA ligand formation if major metal-masking matrix components are removed.

A second mechanism for sample loss is a co-removal process. If the component of interest interacts with the matrix compound that is being depleted, then the component will also be removed from the sample. A similar process occurs in chemistry, known as coprecipitation, where the precipitation of a normally soluble compound can occur when using a "carrier precipitate" – a substance that has a similar crystalline structure and can incorporate the desired chemical species. Although coprecipitation is often undesirable in chemical analysis, in some cases it can be exploited. For example, gravimetric analysis consists of precipitating the analyte and measuring its mass to determine its concentration or purity. Here, coprecipitation can cause problems because any undesired impurities will often also precipitate with the analyte, resulting in excess mass and positive error. This problem may be reduced by slowing the precipitation or heating, and digesting to form larger, more pure particles. The sample may be then redissolved and reprecipitated to form more pure particles.

In the analysis of trace chemical species, however, coprecipitation is often the only way of capturing a particular chemical. Trace chemicals are often too dilute to be precipitated by conventional means, and so are coprecipitated with a carrier.

A chemical example is the separation of francium from other radioactive elements by coprecipitation with cesium perchlorate. In biological samples, a trace component could easily interact with a major matrix biological component, but could be removed along with the major component that is removed by a depletion column.

3.1.2
Roles of Solid-Phase Substrate and Functional Group

Depending on the type of solid-phase extraction, the sample compound may interact with the solid surface of a substrate either directly, or with a functional group bound to the solid substrate. If RNA interacts directly with the solid surface, then it is due to some inherent property or chemical group of the solid-phase substrate.

The functional group provides most solid-phase columns with their selectivity, such selectivity being based on a chemical property of the functional group and of the particular sample molecule that is to be captured. The substrate is what gives the solid phase its form and substance. In this chapter, we will discuss first the various examples of substrate structure, and then the types and manner in which functional groups can be bound. We will then discuss the interactions of sample compounds with the solid phase, with and without functional groups attached, and finally suggest the interaction of sample molecules can be controlled either with functional group or with the substrate directly.

The substrates are chosen based on their uniformity, size, chemical and mechanical stability, and on the ease of attaching the appropriate functional groups to the particle surface. Solid-phase substrate materials are usually based on silica, polymer or agarose core particles. For reasons of inertness and pH stability, a polymer substrate is normally used for the column media in RNA Chromatography, while silica and a polymer substrate is quite commonly used for solid-phase extraction media. Agarose is commonly used for affinity and hybridization interactions.

The goal of solid-phase adsorption is not only to adsorb compounds of interest, but also to release them under specific conditions. Extraction desorption solutions and chromatography mobile phase solutions include salts, buffers, acids, bases, and different organic solvents that are used to control selectivity to retain and to wash the compounds down a column. Each type of separation is based on a particular type of eluent mobile phase that is matched with a column stationary phase, which in turn creates the correct balance of interaction between the column stationary phase and the eluent mobile phase. The selectivity is based on whether the sample compounds prefer to be adsorbed on the column, or to be desorbed and dissolved by the eluent. Specific examples of different combinations of columns and eluents for RNA separation are described in various chapters and sections throughout this book. Step gradients or linear gradients can be used to change the eluent composition as the separation is proceeding. These gradients provide additional control on whether a compound moves down the column, or not. In fact, it is usually more advantageous to keep the sample compound on the column and to wash away all of the impurities; the eluent conditions may then be changed to increase the dissolving power of the eluent (i.e., increase the eluting power of the eluent mobile phase), which desorbs and moves the sample compounds down the column.

3.1.3
Correlation of Interaction Type, Functional Group, and Substrate

Depending on the types of sample compound to be separated, and the desired separation mechanism, different functional groups can be bonded to the core substrate particle, including ion-exchange, reverse-phase, or affinity functional groups. As noted above, there are several types of interaction possible between functional groups and substrates. Functional groups are designed to interact with RNA in a specific manner, while a particular type of solid-phase substrate may often be associated with a particular type of functional group. There are exceptions, however, and new materials and new products will often "break the rules." Hence, the discussion provided below should be taken only as general guidelines.

Some of the most common types of solid-phase materials are listed in Table 3.1. For example, the most common type of RNA (and DNA) chromatography is reverse-phase; the most common substrate is a poly(divinylbenzene) polymer, and the most common functional group is a C18 alkyl group. Interestingly, most commercial reverse-phase columns are silica-based and perform poorly, or not at all, for nucleic acid-type separations.

Table 3.1 Functional groups used for extraction and liquid chromatography.

Interaction type	Functional group	Preferred substrate
Ion pair/reverse phase	Hydrophobic surface	Polymer
Ion exchange	Charged ionic sites	Silica and polymer
Chaotropic	Hydrophilic surface	Silica
Gel filtration	Hydrated porous beads	Water-swollen polymer
Affinity	Selective binding sites	Water-swollen polymer

Most liquid chromatography is performed to separate sample molecules that are relatively small; typical examples are organic acids and alcohols in wine, inorganic ions in drinking water, and surfactants in detergents. While these materials have molecular weights of only a few hundred Daltons, polymers, proteins and nucleic acids by comparison are extremely large, with molecular weights of several thousand Daltons. The requirements to separate large molecules can be much more stringent; indeed, in the case of nucleic acids there are at least three additional requirements:

- It is often necessary to separate and analyze molecules that differ by only one nucleotide in length.
- The separations can be either size-based or sequence-based.

- Because of the high sample load, the separations must be extremely fast and accurate.

These requirements place high standards on the performance of the chromatographic material, which must be highly stable under the conditions used for separation. In addition, temperature stability is required due to the high temperatures of the mobile phase that are used. Although, the pH used for RNA separations is usually about 7, silica-based high-performance liquid chromatography (HPLC) materials are less stable at this pH, especially when there is a high aqueous content of the mobile phase and a high temperature. At a high pH (ca. 10–11), column clearing procedures are also often required to remove sample impurities from the column; notably, polymer columns are more resistant to these cleaning procedures than are silica columns. Historically, silica-based packings have been quite common, mainly because they have been quite successful in the separation of small molecules – almost everybody who has received some training in chromatography prefers silica-based HPLC materials and, indeed, some RNA separations are performed on these materials. However, the large majority of these separations are performed with polymeric materials.

Ion-exchange HPLC is also used mostly for short, single-stranded RNA Chromatographic separations, mainly because a silica substrate might not stand up to the buffers used in ion-exchange chromatography separations. Polymer substrates are also quite commonly used in ion-exchange columns.

The situation changes, however, when extraction materials are considered. Extraction columns are usually single use, and although chemical stability is still important, the stability need not be as longlasting. The pore size, surface area, substrate purity and particle size of silica materials can be finely controlled and also produced at very low cost. Moreover, silica can be functionalized easily with chemical groups (as described below). If silica can be used for an application, it is usually preferred over polymer substrates, with silica-based substrates being quite common for ion-exchange extractions and chaotropic-type extractions.

Agarose, Sepharose (and occasionally cellulose) substrates are used for affinity-type solid-phase extraction. In some separations, selective proteins may capture certain types of RNA, or perhaps a poly T oligo functional group might be used to hybridize messenger RNA. The most striking trait of these types of substrate is their ability to swell in aqueous buffers; in fact, the ratio of aqueous to solid is so high when swollen that the bead is almost entirely water. Because of this aqueous environment found inside these types of substrate, proteins are not easily denatured and more natural buffers can be used. In addition, the materials are fairly pH-stable and a wide pH buffer range can be used.

3.1.4
RNA Structure and Solid Surface Interaction

Both, DNA and RNA are very large molecules composed of a backbone of alternating sugar and phosphate molecules, bonded together in a long, polymeric chain.

The polymer is of varying lengths that, depending on its biological function, range from just a few units up to tens of thousands of units. These varying sizes are known as "nucleic acid fragments". For DNA, the sugar molecule in the backbone is deoxyribose, while for RNA it is ribose. The nucleotide bases are bonded to each sugar molecule, as shown below:

```
 Nucleotide        Nucleotide        Nucleotide        Nucleotide
   base              base              base              base
    |                 |                 |                 |
- sugar - Phosphate - sugar - phosphate - sugar - phosphate - sugar -
              △                 △                 △
```

Here, the symbol △ denotes that the phosphate group has a negative charge. There are four different types of nucleotide base which, in DNA, are adenine (A), thymine (T), cytosine (C), and guanine (G). An RNA nucleic acid molecule is illustrated schematically below, in which the order of nucleic acids is A, uracil (U), C, and G:

```
    A                 U                 C                 G
    |                 |                 |                 |
- sugar - phosphate - sugar - phosphate - sugar - phosphate - sugar -
              △                 △                 △
```

In 1953, Watson and Crick showed that not only is the DNA molecule double-stranded, but that the two strands wrap around each other to form a coil or helix, forming complementary strands. Each nucleotide in one strand is hydrogen-bonded to another nucleotide base in a strand of DNA opposite to the original. This bonding is specific: adenine always bonds to thymine (and *vice versa*), and guanine always bonds to cytosine (and *vice versa*). It should be noted that there are exceptions to this bonding rule, however, which can help give rise to secondary structure in RNA.

Several important similarities and differences exist between RNA and DNA. Like DNA, RNA has a sugar–phosphate backbone to which the nucleotide bases are attached. The bonding occurs across the molecule, leading to a double-stranded molecule as shown below:

```
              ▽                 ▽                 ▽
- sugar - phosphate - sugar - phosphate - sugar - phosphate - sugar -
    U                 A                 G                 C

    A                 U                 C                 G
- sugar - phosphate - sugar - phosphate - sugar - phosphate - sugar -
              △                 △                 △
```

Although RNA can exist in both double-stranded and single-stranded forms, the single-stranded RNA form is more common. RNA is the main genetic material contained in viruses, and is essential for the production of proteins in living organisms. RNA is transported within the cells of living organisms, thus serving as a type of genetic messenger. In this way, information stored in the DNA of the eukaryotic cell is transported by RNA from the nucleus to other parts of the cell, where it is used to help make proteins. It should be noted that not only does the RNA function as a messenger, but that different types of RNA have been shown to function in many other cellular functions, including post-transcription repression, translation repression, catalytic activity, and as adaptors.

The interaction of RNA with a solid-phase reagent occurs though the phosphate group if the interaction is ionic and the solid phase is an anion exchanger. The sequence of the RNA will affect the selectivity; generally, larger RNA molecules have a higher selectivity for a resin because it contains more negatively charged phosphate groups. However, the type of nucleic acid can either increase or decrease the interaction of the phosphate group, such that the separation is not always predictable. While it is possible that a smaller RNA molecule could elute later than a larger molecule, double-stranded RNA generally elutes later than single-stranded RNA because of the greater relative exposure of the phosphate groups.

An ion-pairing reagent such as triethylammonium acetate, when added to the buffers, can convert the RNA molecule to a nonpolar entity, thus making possible a solid-phase interaction with a reverse-phase molecule. Each positive triethylammonium molecule is positively charged, and will pair with the negatively charged phosphate group; in this way, both single-stranded and double-stranded RNAs will become nonpolar. In this case, the double-stranded molecule will also interact more than would the single-stranded molecule.

An additional, very powerful and versatile method is to add a chaotropic reagent (see Section 3.5) to the buffer; for example, the addition of $6\,M$ guanidinium chloride, $6–8\,M$ urea, and $4.5\,M$ lithium perchlorate will cause the DNA to be denatured, thus making possible the interaction with a normal phase surface such as silica. High concentrations of generic salts may also have chaotropic properties.

As we shall see in several examples shown throughout this book, RNA has a structure that can affect the interaction of RNA with a solid-phase surface; sometimes this will benefit the separation, and sometimes harm it. Generally, however, if the RNA molecule has a 3-D structure, then any interaction with the substrate will be decreased and the RNA molecule more easily removed from the solid-phase extraction material.

3.2
The Solid-Phase Substrate and Attachment of Functional Groups

Functional group attachments are generally covalent via covalent bonds, using chemical reactions. Generally, the functional groups are attached during column manufacture, the aim being to obtain identical column selectivities, with an abso-

3 RNA Separation: Substrates, Functional Groups, Mechanisms, and Control

lute minimum lot-to-lot variation. However, if a particular functional group of interest is not available, then it must be synthesized, perhaps employing some of the general synthesis reactions described in the following sections.

3.2.1
Polymeric Resin Substrates

A variety of polymeric substrates can be used in reverse-phase synthesis, including polymers of esters and amides. Polymers are generally prepared via a two-phase polymerization of vinyl monomers, such as styrene, acrylic esters, and vinyl acetate, in a mixture of water, organic crosslinking agent, initiator, and a stabilizing agent. Styrene/divinylbenzene copolymers are probably the most common type of polymer reverse-phase and ion-exchange materials.

A schematic of the polymer is shown in Figure 3.1. Here, the resin is composed of polystyrene or polyethylstyrene, with divinylbenzene added to "crosslink" the various polymer chains in the resin. Crosslinking confers mechanical stability on the polymer beads by interlocking the polymer chains, and because of the high eluent pressures used by chromatographic instruments, a degree of crosslinking of 55–85% by weight of the crosslinking agent is common for chromatographic polymers. Another reason for this level of crosslinking is that the solvents used in chromatography can cause the bead to swell if present, or to shrink if the organic solvent is removed (i.e., if the eluent becomes aqueous). Even with the precaution of using a high degree of crosslinking, a polymer will always swell very slightly in the presence of an organic solvent. As polymer columns are packed with organic solvents, the polymer may be in a slightly swollen state in the column; however, changing the solvent to a completely aqueous mobile phase can lead to a removal of the solvent and cause the bed of the column to shrink and crack. Although the possibility of a cracked bed is less problematic with solid-phase extraction, it should still to be avoided. Generally, the best polymer columns can tolerate 100%

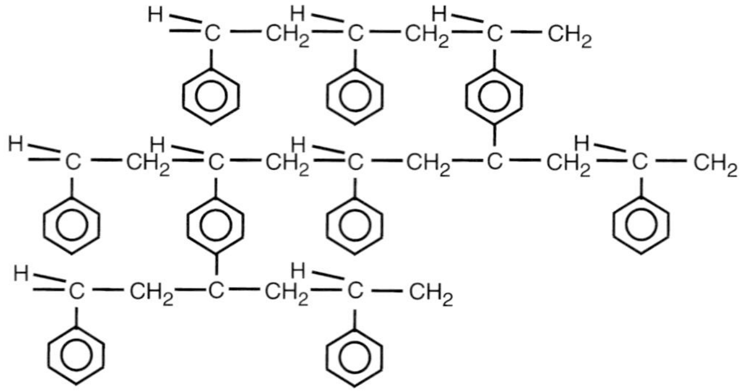

Figure 3.1 Schematic of a styrene–divinylbenzene copolymer. The divinylbenzene "crosslinks" the linear chain of the styrene polymer. A high percentage of divinylbenzene produces a more rigid polymer bead.

aqueous solvents, although it is best to follow manufacturers' instructions with regards to solvent limits.

3.2.2
Porous and Nonporous Polymeric Resins

The "polymer" resins can be classified as either nonporous or porous. The best RNA separations have been performed on nonporous resins, with the nonporous substrates being synthesized by producing a micelle or an emulsion/suspension monomer suspended in water as organic droplets. The monomers are held in suspension in the reaction vessel through rapid, uniform stirring, and the polymerization is initiated by the addition of a free radical catalyst, such as benzoyl peroxide. The mixture is heated either with an oil bath or a heating mantle positioned around the reaction vessel. When the temperature reaches 75 °C, the benzoyl peroxide forms free radicals that will combine with a monomer to form the start of a polymer, with a new free radical on the end. The monomers will be added one by one to the polymer chain until either they have all been used or they can no longer reach the chain end. The solid beads form rapidly (usually within 30 min), but the reaction is heated for up to a further 16 h to ensure that the resin is cured – that is, to ensure that as much monomer as possible has been converted into the polymer. The resultant beads are uniform and solid, and ready to be cleaned.

The resulting size distribution of the solid beads depends on whether a modified micelle process or an emulsion/suspension process has been used. The methods to prepare polymers normally use a combination of processes to form and maintain the monomer spheres before they are polymerized. Although harder to control and more limiting in the final size of the particle, a micelle process produces the most uniform particles. It is usually unnecessary to perform any additional size classification of the beads when this polymerization process is completed.

Bonn and coworkers [1–3] described the most successful modified micelle process used to prepare packing for RNA separations. The process starts with the formation of a suspension of small polystyrene seeds of uniform size; these are produced by introducing a charge to the bead surface during the initial polymerization process. The seeds are then heated or activated in order to accept the transfer of divinylbenzene from an emulsion to the seed particle. As the seeds swell to an appropriate size (2 µm) with the fresh monomer, they become polymerized to form hard solid beads that are then removed, cleaned, and their surfaces modified to the appropriate polarity by adding a C18 alkyl group.

The emulsion/suspension polymerization process is much easier to control and can be used to produce a much wider range of particle-sized materials. Typically, particles with an average size ranging from 2 µm up to 50 µm are produced, although the material produced will tend to have a larger particle size distribution and so must be sized (usually with air classification) before the resin can be used for chromatographic purposes.

The introduction of a detergent into the "monomer mix", followed by rapid stirring in an aqueous buffer, leads to the process of an emulsion/suspension polymerization. The monomer mix includes the monomer, free radical initiator, and an optional pore modifier (poragen) reagent and, when heated, causes the polymerization process to occur, as described above.

Most materials used for nucleic acid separations are nonporous – a term which can be misleading at times. One reason for this definition is that the pores which may be present are too small for the nucleic acid molecules to enter. This is especially true for double-stranded DNA and RNA, where the molecules can be very large. It transpires, however, that most polymers have pores as part of their structure, and hence the term nonporous is in fact used for these "slightly porous" polymers because the pores of the particle matrix do not form part of the separation process. If only part of the RNA molecule were to enter the pore, the kinetic effect would be undesirable because there would be a nonuniform or heterogeneous interaction of the RNA with the beads, and this in turn would lead to broader peaks. Some column materials are said to have "mega" or very large pores where RNA cannot become trapped, but these pores are in effect still part of surface structure and usually do not have any major effect on selectivity. Yet, the material may still be termed "nonporous" if the RNA cannot enter the bead matrix, but only interact with the extended surface of the bead. The situation is somewhat different with short (e.g., 20-mer) single-stranded RNAs, which are small enough to be able to enter the bead matrix while retaining sharp chromatographic peaks.

The use of various ratios of a diluent or poragen (a solvent which is good for the monomer but poor for the polymer into the monomer mix) can permit the pore size of a polymer to be controlled. The resultant resin bead is spherical and comprised of many hard microspheres interspersed with pores and channels. Again, as with the nonporous resins, the polymerization is performed while the monomers are held in a suspension of a polar solvent (usually water). However, the suspended monomer droplets also contain an inert diluent which may be a good solvent for the monomers, but not for the material that has already been polymerized. Generally, the diluents are small organic alkane solvents, but occasionally they may be small linear polymers of controlled and specific molecular weight. Thus, resin beads are formed that contain pools of diluent distributed throughout the bead matrix. When the polymerization is complete, the diluent is washed out of the beads to form the porous structure; the resultant resin beads are rigid and spherical, with a high surface area of $100–400\,m^2\,g^{-1}$. The exact surface area was seen to depend on the type and amount of diluent used, while the pore volume of the bead was directly proportional to the amount of diluent used relative to the amount of monomer used. The pore volume was shown to range from 20% to 80% by volume, with a 50% pore volume being quite common.

Rather than use an inert solvent to precipitate the copolymer and form the pores, the polymerization may be carried out in the presence of an inert solid agent, such as finely divided calcium carbonate, so as to create the voids within the bead; the solid can be extracted from the copolymer (using acid) at a later stage. Both of these polymerization processes create large (albeit probably different) inner pores, with the average pore diameter varying within the range of 20 to 500 Å.

The final resin bead structure of a porous resin contains many hard microspheres interspersed with pores and channels (alternative terms for "porous" include macroporous, macroreticular, and megaporous). However, because each resin bead is composed of thousands of smaller beads (resembling a popcorn ball), the surface area of a macroporous resin will be much higher than that of a nonporous resin. As an example, while (depending on the bead size) a nonporous resin has a (calculated) surface area of less than a few m^2 per gram, macroporous resin surface areas range from 25 up to $800\,m^2\,g^{-1}$. Several reports have been made regarding polymer bead synthesis [4–7].

The chromatographic properties of porous polymers can be easily modified by selecting different vinyl monomers, or by altering the conditions of the chemical reaction of the polymeric beads, including the type and quantity of the poragen and/or the suspension stabilizer. Although styrene-based packings remain the most common, other hydrophilic packings based on polyvinyl alcohol, poly(allylmethacrylate), poly(hydroxethylmethacrylate), and poly(vinylpyridine) are commercially available [8].

3.2.3
Monolith Polymeric Columns

Another approach to producing polymeric columns was introduced by C. Horvath and subsequently expanded by others for the separation of proteins [9, 10]. The approach starts with a fused silica capillary tubing, the walls of which have been conditioned to accept the attachment of monomers during a polymerization process. A mixture of monomers, poragen and catalyst is then filled into the tube, and the ends of the capillary are sealed. The mixture is heated to start the polymerization process, such that a solid polymer and pore structure is produced inside the capillary, including attachment to the walls. The ends are cut and the capillary is then cleaned with solvent, making it ready for chromatographic use. This type of column is known as a "monolith" because the polymer structure extends throughout the column, and no frits are required to keep the column packing material in place.

A pore structure which extends throughout the column allows the solvent to flow through the column, and also allows interaction of the sample with the monolith stationary phase. As with conventional materials, the pores for monolith columns are formed from the addition of inert poragens that have been added to the monomer mix. The resultant pore structure of the monolith actually is quite similar to that of conventional packing materials, with the pores that interconnect and extend through the monolith (the "through pores") being analogous to the interstitial spaces between packing particles. The monolith polymer matrix itself also contains a pore structure which is similar in structure and performance in conventional materials to the pore structure of the packing beads. For RNA Chromatography, these monolith matrix pores are quite small and do not allow penetration of the nucleic acids into the resin matrix.

3.2.4
Functionalization of the Polymer

The most common method of controlling the surface properties is to cover the surface of the resin bead with chemical functional groups; in the case of reverse-phase chromatography, C18 alkyl groups were the most frequently attached. One reason why poly(styrene/divinylbenzene) is so popular for this process is because the aromatic group of the substrate can be reacted in many different ways to attach the alkyl group. An example is shown in Equation 3.1:

$$\text{Resin}-\underset{}{\bigcirc} \xrightarrow[\text{SnCl}_4]{\text{ClCH}_2(\text{CH}_2)_{16}\text{CH}_3} \text{Resin}-\underset{}{\bigcirc}-\text{CH}_2(\text{CH}_2)_{16}\text{CH}_3 \quad (3.1)$$

Other functional groups that can also be attached include cyano groups, esters, and amides, all of which are more polar, although the surface is still hydrophobic and might potentially serve as beads for reverse-phase separations. Nonetheless, alkyl groups remain the most popular for reverse-phase procedures, with polymer resins having this functional group demonstrating the highest efficiencies of separation.

Ion exchangers are prepared in a very similar manner; the treatment of a resin with chloromethylmethylether and a tin catalyst is shown in Equation 3.2. The second step involves treatment with an amine to form an anion exchanger; different types of amine can be used here, depending on the selectivity desired for the final product. For example, trimethylamine can be used to produce a strong base–anion exchanger (Equation 3.3), and diethylamine to produce a weak base–anion exchanger (Equation 3.4). The diethylaminoethyl anion exchanger (DEAE) is a positively charged resin that is commonly used in protein and nucleic acid separation and purification. As the weak base–anion exchangers are tertiary amines, the pH of the buffer must protonate the group before it can act as an anion exchanger. The DEAE groups have a pK_a of 11, which means that the pH must be below this value before the DEAE functional group can retain negatively charged sample compounds.

$$\text{Resin}-\underset{}{\bigcirc} \xrightarrow[\text{SnCl}_4]{\text{ClCH}_2\text{OCH}_3} \text{Resin}-\underset{}{\bigcirc}-\text{CH}_2\text{Cl} \quad (3.2)$$

$$\text{Resin}-\underset{}{\bigcirc}-\text{CH}_2\text{Cl} \xrightarrow{\text{N(CH}_3)_3} \text{Resin}-\underset{}{\bigcirc}-\text{CH}_2\overset{+}{\text{N}}(\text{CH}_3)_3\ \text{Cl}^- \quad (3.3)$$

or

$$\text{Resin}-\underset{}{\bigcirc}-\text{CH}_2\text{Cl} \xrightarrow{\text{N(CH}_2\text{CH}_2\text{OH})_2} \text{Resin}-\underset{}{\bigcirc}-\text{CH}_2\overset{+}{\text{N}}(\text{CH}_3)_3\ \text{H Cl}^- \quad (3.4)$$

3.2 The Solid-Phase Substrate and Attachment of Functional Groups

A wide variety of resins based on polyacrylate polymers have been produced for use in chromatography. The type known as HEMA (a macroporous copolymer of 2-hydroxyethyl methylmethacrylate and ethylene dimethacrylate) has been used extensively in both reverse-phase and ion-exchange chromatography. HEMA is highly crosslinked so as to produce a polymeric matrix with high chemical and physical stability; the structure is shown in Figure 3.2. Esters are composed of organic acids and alcohols bonded together, and can easily be hydrolyzed or become unbound in aqueous solution (especially at high pH). The highly hindered structure of pivalic acid is one of the most stable and least hydrolyzable esters known, and allows the HEMA stationary phase to be used with a variety of eluents in the pH range of 2 to 12. While these may be less rugged than a poly(styrene/divinylbenzene)-based substrate, the pH stability was still considerably better than that of the silica-based substrates described in Section 3.2.5. The excess hydroxyl groups on the HEMA matrix are rather versatile, having the ability to increase the hydrophilicity of the material and improve interactions with polar sample compounds. The groups can also provide a reaction site upon which more traditional reverse-phase groups and anion-exchange sites can be attached.

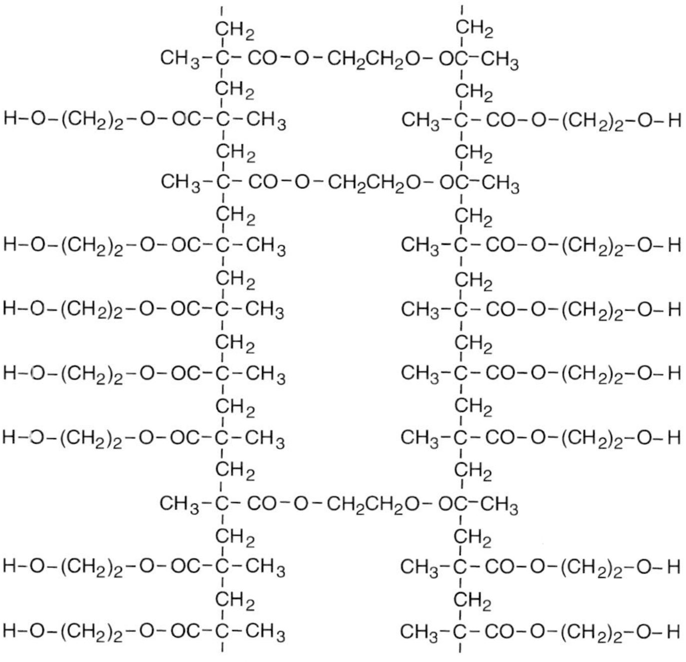

$$
\begin{array}{c}
\text{CH}_2 \\
\text{CH}_3-\text{C}-\text{CO}-\text{O}-\text{CH}_2\text{CH}_2\text{O}-\text{OC}-\text{CH}_3 \\
\text{CH}_2
\end{array}
$$

[chemical structure diagram of HEMA polymer matrix showing repeating units of CH₃–C(CH₂)(CH₂)–CO–O–CH₂CH₂O–OC–C(CH₂)(CH₂)–CH₃ crosslinks and pendant H–O–(CH₂)₂–O–OC–C(CH₂)(CH₂)–CH₃ and CH₃–C(CH₂)(CH₂)–CO–O–(CH₂)₂–O–H side groups]

Figure 3.2 The structure of HEMA. Most esters are easily hydrolyzed or unbound in aqueous solution (especially at high pH). The hindered structure of pivalic acid is one of the most stable and least hydrolyzable esters known, which allows the HEMA stationary phase to be used with a variety of eluents in the pH range 2 to 12. The hydroxyethyl group allows the chemical attachment of organic functional groups.

The Dionex Co., of Sunnyvale, CA, USA, has developed a number of latex-coated nonporous materials. Several of these anion-exchange resins have a surface layer of quaternary ammonium latex on a surface-sulfonated substrate. The schematic for these anion-exchangers is:

$$\text{Resin} - SO_3^- N^+ R_3 - \text{latex} - N^+ R_3 \tag{3.5}$$

Some of the anion-exchange sites in the latex are used to attach it electrostatically to the resin substrate, leaving the outside anion-exchange sites of the latex free for use in anion-exchange chromatography.

Whether a particular material can be used for the separation of nucleic acids depends on a number of factors, including pore size, ion-exchange capacity, stability, and so on. Hence, it is important to follow the manufacturers' recommendations as to what materials are best suited for these procedures.

3.2.5
Silica–Glass-Based Substrates

Silica and glass substrates are preferred when pH stability is not an issue. As stated before, most recent liquid nucleic acid separations have been performed on polymer-based substrates, on the basis of improved stability and separating power. Nonetheless, silica-based substrates remain quite common in the traditional solid-phase extraction of RNA. A brief description of silica substrates is now provided.

The production of silica-based materials for reverse-phase separations follows a similar path as for polymers. First, the substrate is produced of a suitable particle size, surface area, and pore size, after which the silica substrate is reacted to attach a functional group with the desired properties of selectivity.

The first silica chromatographic-based materials were prepared using a sol–gel procedure. For this, a hydrosol is initially prepared by the addition of sodium tetraethylsilate to an acidic aqueous solution. An initial hydrolysis occurs, followed spontaneously by condensation to poly(silicic acid). As a result, an insoluble cake is formed that is eventually dehydrated to form a hard, porous material that can be milled and then sized in an air stream to collect the desired particle size range.

The particles produced by this method are irregular and not optimum; however, spherical microparticles are more desirable because of their improved performance, as measured by the van Deemter plot (see Appendix 1). In spite of the fact that spherical silica substrates are more difficult to manufacture, most commercial silica-based particles are in fact spherical.

Spherical silica can be prepared by several methods [11, 12]. In one such method, silica hydrogel beads are prepared by the emulsification of a silicic acid sol in an immiscible organic liquid. The hydrogel is then dispersed into small droplets in the organic liquid, and the temperature, pH, and electrolyte concentration are each adjusted to promote solidification. Over time, the liquid droplets become more viscous and solidify into bead-type particles that are then dehydrated to form porous spherical silica beads. The particle size and pore diameter can be controlled by the size and extent of hydration of the polysilicic acid used to initiate the process.

An alternative approach is based on the agglutination of a silica sol by coacervation (which means to "heap up and harvest"). For this, urea and formaldehyde are polymerized at low pH in the presence of colloidal poly(silicic acid). Subsequently, coacervated liquid droplets with a spherical shape are formed and settle from the reaction medium. In time, the beads solidify and are harvested and dried, after which the polymer binder is burned out to form the pores of the spherical particles [13, 14].

Among nonporous silica which have also been produced, the earliest materials – known as "pellicular" – had quite large particle sizes and were considered to be less efficient for separations when compared to porous silica materials. More recently produced materials have a quite small particle size (some as low as 1 μm) and can be used to perform separations quite rapidly. Although very high backpressures may result from these extremely small-particle packings, this can be controlled by using very short columns of perhaps only 30 mm [15–17].

3.2.6
Functionalization of Silica

In order to operate as a reverse-phase material, the inorganic silica substrate must have a nonpolar molecule bonded to its surface. These nonpolar molecules bound to the inorganic substrate may be either organic polymer or long-chain, C1 to C24 hydrocarbon groups. The silica surface consists of a network of silanol groups and siloxane groups, as shown in Figure 3.3. The bonded phases are prepared by reaction of the surface silanol groups and reactive organosilanes to form siloxane bonds, with the organosilane imparting the desired surface effect. A typical reaction is shown in Equation 3.6:

$$\text{Substrate} - \text{Si} - \text{OH} + \text{Cl Si (CH}_3\text{)}_2\text{R} \longrightarrow \text{Substrate} - \text{Si} - \text{O} - \text{Si (CH}_3\text{)}_2\text{R} \qquad (3.6)$$

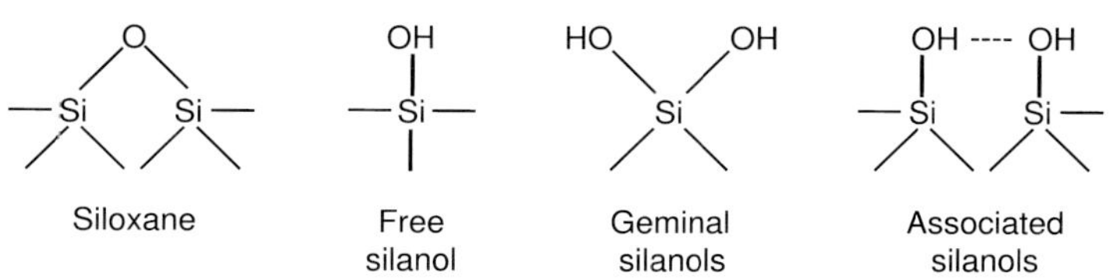

Siloxane   Free silanol   Geminal silanols   Associated silanols

Figure 3.3 Types of silanol groups that may be present on a silica surface. The chemical attachment of functional groups to silica is performed using chloro- or methoxy-silane compounds containing the alkyl or ion exchange group reacted to the silanol sites.

where R is the alkyl group of any desired length (usually C18) or polarity. Initially, the reaction may be stoichiometric, but as the surface coverage approaches completion it will become very slow. The typical overall reaction times tend to be long (12–72 h), with temperatures of approximately 100 °C.

Unfortunately, the reaction is never complete and residual silanol groups always remain, this being the primary reason why silica-based materials are more unstable under the conditions used for RNA separations. As many undesirable chromatographic properties of bonded phases are associated with the presence of these accessible silanol groups, further investigations are required to reduce their presence, or at least the effect of their presence. One way to accomplish this is by performing a secondary reaction with chlorotrimethylsilane, which is smaller than the original organosilane reagent and can access the unreacted silanol sites. The replacement of accessible silanol groups in a bonded phase by trimethylsilyl group is generally referred to as "endcapping".

Other methods used to reduce the effect of silanol groups include polymer encapsulation, which can be achieved by coating the silica substrate with a thin film of prepolymer; the latter is then crosslinked to form an immobilized skin over the silica surface. The prepolymer may be simply coated onto the substrate, or it may be bonded through the silanol groups [18–21]. A common type of silica anion exchanger has the structure:

$$\text{Substrate} - O - Si - CH_2 - \underset{}{\bigcirc} - CH_2N^+R_3 \ A^-$$

Compared to organic polymers, silica-based ion-exchangers have the advantages of higher chromatographic efficiency and greater mechanical stability. In general, no problems due to swelling or shrinking are encountered, even if an organic solvent is added to the eluent. One disadvantage of silica materials, however, is their limited stability at lower pH values, and especially in alkaline solutions. Consequently, a fairly narrow pH range of 2 to 8 is recommended for these reactions.

Silica-based anion-exchangers are available from several manufacturers; common trade names include Vydac® (Separations Group, Hesperia, CA, USA), TSK Gel® (Toyo Soda, Tokyo, Japan), and Nucleosil® (Machery & Nagel, Düren, Germany). As noted above, there are two distinct types of silica resin:

- Totally porous resins have a quaternary ammonium functional group which is chemically attached. Their particle size is in the range of 3 μm to 10 μm, with typical exchange capacities of 0.1 to 0.3 mEq g^{-1}.

- Nonporous resins, sometimes called pellicular resins, have a larger particle size and are covered with a thin layer of a polymer with quaternary ammonium groups. For example, Zipax SAX® is covered with a layer of lauryl methacrylate which is 1–3 μm thick.

3.2.7
Agarose and Cellulose Affinity Substrates

Agarose is a polysaccharide with a galactose-based backbone, while cellulose is a polysaccharide linear chain D-glucose. The neutral charge and low degree of chemical complexity of both agarose and cellulose makes them ideal substrates for separating proteins and nucleic acids.

Agarose is by far the most common substrate; beads prepared from purified agarose have a relatively large pore size, which makes them useful for the size-separation of large molecules. Crosslinking of the agarose chains to form beads can be achieved with reagents such as epichlorohydrin or dibromopropanol.

The hydroxyl groups of agarose and cellulose can be partially or fully reacted with a variety of reagents to add functional groups with useful extractive properties. As an example, cyanogen bromide is often used to immobilize proteins by coupling them to reagents such as agarose for affinity chromatography. Activation with cyanogen bromide is performed under mild pH conditions, and is probably the most common method used for preparing affinity gels. Here, the cyanogen bromide reacts with the hydroxyl groups on agarose to form cyanate esters and imidocarbonates; these groups are then reacted with primary amines in order to couple the protein onto the agarose matrix. One disadvantage of cyanogen bromide is the sensitivity to oxidation. Such activation may also involve attaching a ligand to agarose via an isourea bond, which is positively charged at neutral pH, and is thus unstable. Consequently, isourea derivatives may act as weak anion-exchangers in the final affinity product.

3.2.8
Dextran and Polyacrylamide Gel Filtration Substrates

Gel-filtration chromatography is also referred to as gel permeation, desalting, and buffer exchange. The substrates usually do not have functional groups attached, but rather are aqueous swollen gels. The most popular material for RNA separation appears to be the polysaccharide dextran, a complex, branched polysaccharide which is composed of many glucose molecules joined into chains of varying lengths (from 10 to 150 kDa). An alternative material, polyacrylamide, is a polymer ($-CH_2CHCONH_2-$) formed from acrylamide monomer subunits. A list of gel filtration materials which have smaller pores and are suitable for RNA separations is provided in Table 3.2.

The pore sizes of both affinity and gel filtration beads can be controlled by the degree of polymer crosslinking, which is the chemical bonding of the linear polymers that make up the beads. As the degree of crosslinking is increased, the polymer bead is less able to swell and so the pore size is decreased. Since only those molecules that can enter the pores are retained, the higher-crosslinked substrates will be capable of separating only smaller sample compounds.

Table 3.2 Gel filtration materials and separation size range.

Name, material	Particle size (µm)	Separation size range (Da)
Sephadex, dextran		
G-10	40–120	up to 700
G-25 medium	50–150	100–5 000
G-25 medium	50–150	500–10 000
Bio-Gel P, polyacrylamide		
P-6 DG	90–180	1 000–6 000
P-10 medium	90–180	1 500–20 000
Superdex, agarose / dextran		
Superdex 30 prep grade	22–44	up to 10 000
Superdex 75 prep grade	22–44	500–30 000
Superdex 75	11–15	500–30 000

3.3
Reverse-Phase Ion-Pairing Separation Mechanism

In order for RNA to be separated by the reverse-phase column, it must first interact with the stationary phase. The stationary phase is nonionic and hydrophobic, while the polymeric RNA molecule is ionic due to the phosphate backbone. RNA cannot interact with the solid-phase surface and so must become nonpolar in order to develop an extraction or chromatographic procedure. To accomplish this, an organic cation is added to the sample and the buffer solutions to render the RNA nonpolar such that it will adsorb to the column.

For this purpose, the triethylammonium cation is a useful ion pairing cation, and is prepared by adding acetic acid to triethylamine to create a final solution with a pH of 7:

$$N(CH_2CH_3)_3 + HAc \rightarrow HN^+(CH_2CH_3)_3 + Ac^- \tag{3.7}$$

where Ac^- denotes acetate. The ion-pairing solution is added to the RNA sample to form the ion pair, as shown in Equation 3.8:

$$HN^+(CH_2CH_3)_3 + RNA\,PO_4^- \rightarrow RNA\,PO_4^- N^+(CH_2CH_3)_3 \tag{3.8}$$

In order for the mechanism to operate properly, the eluent buffers must also contain the ion-pairing reagent. An organic solvent such as acetonitrile is then used to elute the RNA from the column.

The separation process is shown in Figure 3.4. On starting with a gradient of acetonitrile, the smaller RNA molecules are first desorbed, travel down the column, and are detected. As the organic solvent content of the eluent is then raised, increasingly larger RNA molecules are desorbed and also travel down the column.

3.3 Reverse-Phase Ion-Pairing Separation Mechanism

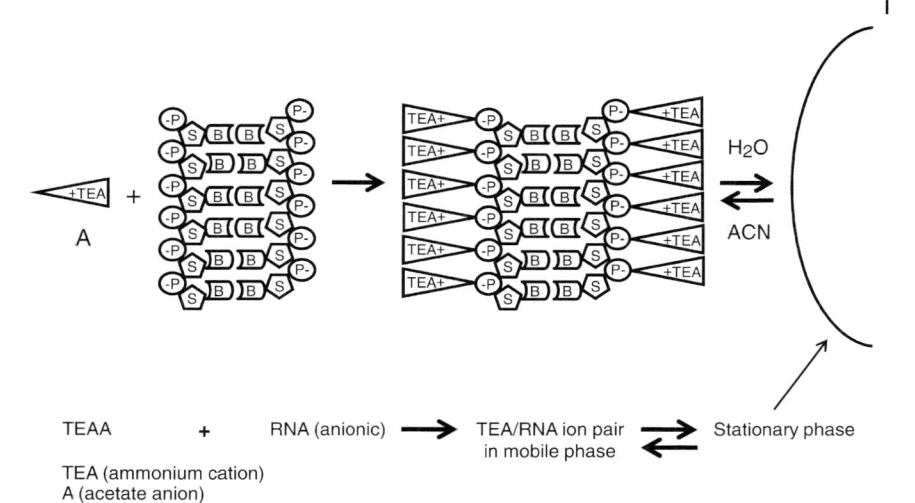

TEAA + RNA (anionic) → TEA/RNA ion pair in mobile phase ⇌ Stationary phase

TEA (ammonium cation)
A (acetate anion)

Figure 3.4 The ion-pairing reverse-phase process. TEA = triethylammonium; A = acetate; P = phosphate; S = sugar; B = base. The ion-pairing reagent is prepared by mixing triethylamine with acetic acid to form triethylammonium acetate (TEAA). The negatively charged nucleic acid is combined with the positively charged triethylammonium ion to form the neutral ion pair. The ion pair can interact with the surface of the column packing (the stationary phase). As the water content of the mobile phase increases, the equilibrium or adsorption to the neutral nonporous surface increases. As the acetonitrile concentration of the mobile phase increases, the equilibrium or adsorption to the surface decreases.

When all the fragments of interest have been eluted, the column is cleaned of residual material with a final gradient concentration of high acetonitrile content. Subsequently, the eluent is returned to the original concentration and the column is either conditioned for the next run, or disposed of.

The concentration of the injection solvent (i.e., the solvent containing the sample) must be less than that of the eluent; otherwise, the chromatographic process will begin prematurely. For example, if a separation is started at 10% acetonitrile, but the sample acetonitrile solvent is 15%, some of the sample will begin to travel down the column before the gradient is initiated. This will lead to broad, nonreproducible peaks, or may even cause the sample to be washed through the column so that any separation is completely prevented.

Double-stranded RNA may be separated on the basis of fragment size; the sequence of the fragment does not contribute to the separation. Single-stranded RNA can also be separated on the basis of size, but the separations are not completely sequence-independent. As these single-stranded molecules contain only half as many phosphodiester groups as double-stranded RNA, they also contain only half as many nonpolar, ion-pairing molecules, and therefore the polarity effects due to the single-stranded nucleic acid backbone will have a greater effect on retention. The retention times of single-stranded nucleic acids are shorter, and less acetonitrile is required to elute them from the column. In addition, fragments containing more C and G nucleotides will elute earlier in the chromatogram than fragments containing more A and U nucleotides. This effect is even greater if the C and G nucleotides are at the end of the fragment. This sequence-dependent effect can be lessened for single-stranded RNA if a more nonpolar (hydrophobic) ion-pairing reagent such as tetrabutylammonium bromide [1–3] or N-hexylammonium acetate ion [22] is employed instead of the more commonly used triethylammonium acetate.

The following experimental parameters can be adjusted to obtain satisfactory conditions for a separation:

- The type of stationary phase: Both, a greater contact surface and a more hydrophobic surface will increase the retention.
- Type of pairing reagent: A more nonpolar reagent shifts the equilibrium towards the surface association.
- Concentration of pairing reagent: A higher concentration shifts the equilibrium towards the surface association.
- Type and concentration of organic solvent (organic modifier): Increasing concentrations shift the equilibrium away from the surface.

The mechanism of what is termed "ion-pairing on a reverse-phase substrate" has been the subject of a considerable amount of investigation. Horvath *et al.* demonstrated the practicality of this approach for chromatography [23], by proposing an ion-pair mechanism and deducing a number of ion-pair-formation constants. Kraak, Jonker and Huber used anionic surfactants in conjunction with a bonded-phase silica column and an organic–aqueous mobile phase for the separation of amino acids [24]. Here, a comprehensive study was made of the parameters, including the generation of gradients. Kissinger argued that an ion-pair mechanism was incorrect [25], but that the pairing reagent would partition strongly onto the stationary phase, thus modifying its surface charge, which in turn implied an ion-exchange mechanism. This interpretation would appear to be valid when the pairing reagent was very strongly adsorbed onto the stationary phase surface. The situation is much like using a reversed-phase HPLC column packing with a permanent coating of a surfactant for ion chromatography. Bidlingmeyer and coworkers also argued that the mechanism was a combination of ion exchange and adsorption [26].

In both, RNA extraction and chromatography, it appears that the separation is based on a desorption of the ion pair from the column as the acetonitrile concentration reaches a certain level. The reasons for stating this are based on certain observations, including:

- When the mobile phase composition is varied for isocratic elution (i.e., no gradient used), the sample is found never to migrate normally, but is either completely retained or completely eluted with the void volume. In other words, depending on the concentration of acetonitrile in the eluent, the RNA will either completely stick to the column, or will not interact at all and travel with the injection solvent plug (into the mobile phase) and through the column.

- When the column length or size is varied and a step elution gradient is used, the resulting separation changes very little. A 3 cm column will give virtually the same separation as a 10 cm column. There is little difference in the retention times and total separation time when samples are run under identical conditions with columns of two different lengths.

- A particular size fragment of RNA will elute at a particular concentration of acetonitrile, regardless of the slope of the gradient (i.e., how quickly the gradient is generated).

Figures 3.5a–c show how the type of gradient can affect how RNA fragments travel through a chromatographic column and are separated. Both, the separation mechanisms of desorption (i.e., release of the fragment from the stationary phase surface) and partitioning (i.e., the adsorption and desorption of the fragment on the stationary phase surface) can occur under RNA Chromatography conditions.

When an eluent gradient is used (Figure 3.5a), only the top portion of the column is used in the separation mechanism. All of the fragments are absorbed with the sample injection. Then, as the gradient is increased each fragment is in turn effectively desorbed from the column, starting with the smaller fragments and continuing until the last (largest) fragment is eluted. At the moment when any particular fragment is no longer interacting with the stationary phase, it will travel through the column with the same linear velocity as the eluent.

In Figure 3.5b, it can be seen that lowering the slope of the gradient will allow more of the column to be used in the separation process. With a shallow gradient, the conditions for complete desorption for a particular fragment are reached over a period of time, during which there is a partitioning of the fragment with the stationary phase. As the gradient is increased, the fragment will eventually cease interaction with the column, and travel through the column at the same linear velocity as the eluent. A higher resolution of the peaks is achieved with a more shallow eluent gradient.

In isocratic elution (Figure 3.5c), the fragment is interacting (partitioning) with the column and eluent. The fragment travels through the column at a constant velocity, but lower than the linear velocity of the eluent. Although this type of chromatography achieves the highest resolution of peaks, it can be difficult to determine the correct eluent concentration for the desired separation of fragments.

This same mechanism of separation is present whether a gradient of acetonitrile is used in ion-pairing reverse-phase chromatography or a gradient of increasing ion (e.g., chloride) concentration is used in ion-exchange chromatography (see below).

3.4
Ion-Exchange Separation Mechanism

Ion-exchange chromatography has a long history, with glass, clays, minerals, and naturally occurring organic substances with ion-exchange properties having been

(a) Gradient elution

Tightly held peaks →

Just released peak coming up rapidly to the eluent velocity →

Early released peak at eluent velocity →

Good resolution

(b) Shallow gradient elution

Tightly held peaks →

Just released peak coming up slowly to the eluent velocity →

Early released peak at eluent velocity →

Better resolution but longer separation time

(c) Isocratic gradient elution

Tightly held peaks →

Partitioning peak moving through column at speed lower than eluent velocity →

Early partitioning peak moving at speed lower than eluent velocity →

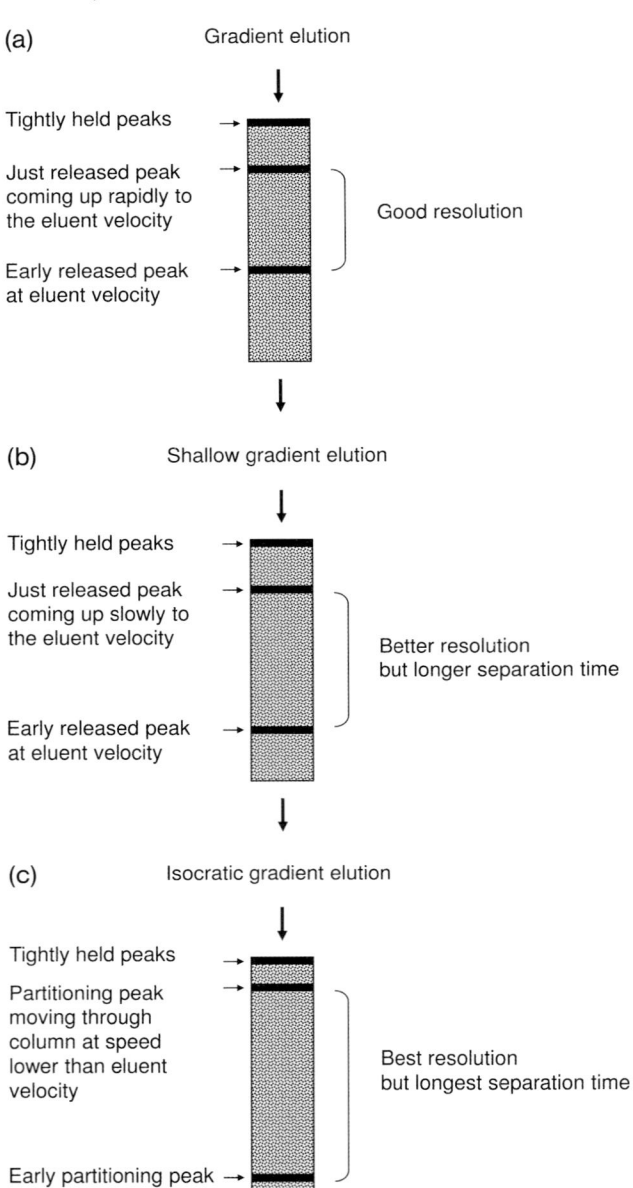

Best resolution but longest separation time

Figure 3.5 (caption see page 59)

exploited for commercial purposes over the years. Yet, the intense use of ion exchange did not really begin until the 1940s, with the development of synthetic, polymeric ion exchangers. Today, ion-exchange chromatography is used extensively for purifying useable quantities of materials, especially in the production of rare earth metals, which are similar chemically and very difficult to purify by any other method. Ion-exchange is also popular for the isolation of preparative amounts of proteins, mainly because the activity of the protein is retained with this procedure.

The analytical uses of ion exchange include the analysis of small ions such as chloride, sulfate and others in drinking water, power plant water, plating baths, food, and beverage products. Anion exchangers with diethylaminoethyl groups were used by Y. Kato *et al.* [27] to separate polynucleotide fragments, with the most important application of anion exchange for RNA being the analysis and purification of short, single-stranded molecules.

One important property associated with anion-exchange separations of double-stranded and single-stranded nucleic acids is the differing retention behaviors of the GC, AT, and AU base pairs. Anion exchange selectivity is known to be based on molecular size and sequence, yet W. Bloch [28] demonstrated that, to a certain extent, a length-relevant separation of double-stranded RNA fragments was possible on nonporous anion exchangers with mobile phases containing tetramethylammonium chloride (TMAC). Another important property of the anion-exchange methodology is the need to use salts and buffers for elution.

Ion exchangers may involve either anions or cations. A typical anion exchanger is a solid particulate material with positively charged functional groups arranged in a manner such that they interact with ions in the surrounding liquid phase. The single phosphate groups on each nucleotide each contribute one negative charge to the total molecular charge and, as nucleic acids are negatively charged, they can be separated using an anion exchanger.

The basis for separation in ion exchange lies with differences in the exchange equilibrium between the various sample anions and the eluent anion. In order to perform a separation, the eluent is passed through the column until all of the ion exchange sites are in the eluent form. For example, if a sodium chloride eluent is equilibrated with the column, then all of the anion exchange sites will have eluent

Figure 3.5 (a) When an eluent gradient is used, only the top portion of the column is used and each fragment is effectively desorbed from the column when the eluent concentration is reached that effectively desorbs each fragment. At this point, the fragment is no longer interacting with the stationary phase, and is traveling through the column at the same linear velocity as the eluent. (b) Lowering the slope of the gradient will allow more of the column to be used in the separation process because the conditions for complete desorption for a particular fragment are reached later. A higher resolution of the peaks is achieved with a shallower eluent gradient. (c) In isocratic elution, the fragment partitions between the column and eluent phases. The fragment is traveling through the column at a constant but lower velocity than the linear velocity of the eluent. The velocity of the fragment will approach the velocity of the eluent as the eluent driving solvent or ion is increased. Isocratic elution gives the highest resolution, but is the most difficult to perform.

chloride anion E⁻ associated with them. Another way of saying this is that the anion exchanger is in the E⁻ form or Cl⁻ form. The time required to achieve this may be several minutes, depending on the type of resin and eluent used.

When an analytical sample containing fragments F_1, F_2, F_3, ... F_i is introduced to the top of the column, each of the fragments undergoes an ion-exchange with the exchange sites near the top of the chromatographic column:

$$F_1 + x\text{Resin-N}^+R_3E^- \rightleftarrows x\text{Resin-N}^+R_3F_1 + xE^- \tag{3.9a}$$

$$F_2 + y\text{Resin-N}^+R_3E^- \rightleftarrows y\text{Resin-N}^+R_3F_2 + yE^- \tag{3.9b}$$

$$F_3 + z\text{Resin-N}^+R_3E^- \rightleftarrows z\text{Resin-N}^+R_3F_3 + zE^- \tag{3.9c}$$

and so on.

Each fragment contains one phosphate group or one negative charge for every nucleotide of the fragment. Every time a fragment exchanges with the anion exchanger, x eluent ions are displaced. Thus, as the fragment size is increased, the number of sites of attachment increase and the fragment is held tighter by the column.

The general principles for separation are perhaps best illustrated by a specific example. Suppose that a 20-mer oligonucleotide and 19-mer failure sequence are to be separated on a higher performance anion-exchange column with a $1.0\,M$ NaCl eluent mobile phase.

In the column equilibration step, the column packed with solid anion-exchange particles (designated as N^+Cl^-) is washed continuously with the NaCl eluent to convert the ion exchanger completely to the Cl⁻ form. In the sample injection step, a small volume of sample is injected into the ion-exchange column, after which an ion-exchange equilibrium occurs in a fairly narrow zone near the top of the column:

$$19\,N^+Cl^- + 19F^- \rightleftarrows 19\,N^+F^- + 19Cl^- \tag{3.10a}$$

$$20\,N^+Cl^- + 20F^- \rightleftarrows 20\,N^+F^- + 20Cl^- \tag{3.10b}$$

where the 19-mer failure sequence is denoted as 19F⁻ and the 20-mer fragment as 20F⁻. Each fragment displaces 19 and 20 Cl⁻ anions, respectively, when the fragment exchanges with the anion exchanger.

In isocratic elution, a constant eluent concentration is pumped through the column. Using the example of a $1\,M$ NaCl eluent through the column results in multiple ion-exchange equilibria along the column in which the sample fragments (19F⁻ and 20F⁻) and eluent ions (Cl⁻) compete for ion-exchange sites next to the N^+ groups. The net result is that both 19F⁻ and 20F⁻ move down the column, although because 20F⁻ has a greater affinity for the N^+ sites than 19F⁻, it will move at a slower rate. Due to these differences in the rate of movement, 20F⁻ and 19F⁻ are gradually resolved into separate zones or bands.

Unfortunately, isocratic elution does not work well for the ion exchange of nucleic acids, as the ion-exchange binding increases rapidly with increasing length of the nucleic acid. Rather, a gradient must normally be used to elute nucleic acids of different lengths in the same chromatogram. The ion-exchange mechanism is similar to that described in Figure 3.4, where the fragment interaction with the column becomes less as the steepness of the gradient is increased. As the concentration of the driving anion, chloride, is increased, then the equilibrium will be shifted and the fragment released from the column.

3.5
Chaotropic Denaturing Interaction Mechanism

The effectiveness of various chaotropic salts on the extraction follows the chaotropic series; $ClCCOONa = NaClO > NaBr > NaCl$. Guanidinium chloride has been the chaotropic salt of choice since the late 1960s [29], and is used to maintain nucleic acid solubility and secondary structure at temperatures up to 70 °C, while the proteins are completely denatured at even very low temperatures. Additionally, high concentrations of guanidinium chloride do not impart any electrostatic forces on the nucleic acids, which prevents the helices from melting apart into single-stranded molecules. When using UV spectrometry to quantify nucleic acid concentrations after purification, guanidinium chloride will not interfere with the spectrum because it is transparent in the UV range. When extracting RNA from an RNA – protein complex, guanidinium is especially useful because of its efficacy in denaturing proteins – a fact not lost on investigators concerned with the stability of RNA following cell lysis, when the RNA is exposed to cellular RNases. Hence, guanidinium chloride is used routinely today to denature the cellular RNases after cell lysis.

When separating nucleoprotein complexes, a phase separation is generally used. For this, the nucleoprotein complexes (e.g., ribosomes) are buffered in 50 mM Tris (pH 7.6) that contains salts to maintain the RNA structure (e.g., 1 mM $MgCl_2$ and 25 mM KCl). The solution is adjusted to at least 4 M guanidinium chloride and incubated on ice. Although increasing the concentration of guanidinium chloride to 6 M will help to maintain the RNA secondary structure, a 4 M concentration cooled to 0 °C provides RNA stability similar to that of RNA buffered in neutral phosphate buffer at room temperature. In addition, as the melting of RNA is reversible, it is not critical to maintain precise temperatures, as long as the temperature does not become so high as to cause hydrolytic cleavage of the RNA. Following the method of Chomczynski and Sacchi, an equal volume of a phenol–chloroform mixture is added to the solution [30]. These authors found that the RNA – protein separation could be achieved by low-speed rather than ultracentrifugation (as was the prevailing thought at the time), thereby relieving a significant bottleneck in the procedure. Following extraction by phenol, the RNA remains in the aqueous phase, and may then be recovered by the addition of salt (sodium acetate, pH 5) and 3 volumes of −20 °C ethanol to precipitate the RNA. Incubation at low temperatures, followed by a 10 min spin in a microfuge tube, will pellet the RNA, which can then be resuspended in buffer.

3.6
Hybridization

Nucleic acid polymers have unique hydrogen-bonding capabilities and a propensity for forming duplex molecules. The length of the duplex confers both specificity and affinity. The energetic contribution to the stability of the helix derives from the base-stacking interactions of the π orbitals of the aromatic nucleotides, as well as from the hydrogen-bonding interactions. If the sequence of a specific RNA is known, it can be purified from complex samples by way of hybridization. Solid-phase substrates can be functionalized with short RNA or DNA sequences which are complementary to the target RNA. In practice, the length of the functionalized RNA is greater than 15 nucleotides, with the RNA sample being applied to the column under neutral pH and physiological salt conditions. The target RNA will interact with the functional groups, and sample contaminants are washed away; an elution with salt or varying pH will then release the pure target RNA. Notably, this approach can be applied equally well to extraction methods.

In eukaryotes, mRNAs are modified with a string of adenines at the 3′ end of the molecule, known as the poly(A) tail, the length of which is in the range of 20 to 250 nucleotides. Hybridization methods can be used to extract these mRNA molecules from the remainder of the cell by using poly (dT) oligos, and many kits using beads functionalized with poly (dT) oligos of 30 nucleotides in length are available commercially. These kits allow for the capture and purification of mRNAs for subsequent expression analysis. Moreover, when coupled with microarray technology, it is possible to study changes in gene expression patterns between different sets of samples.

Microarray hybridization techniques allow for the simultaneous measurement of all the expressed genes of a cell. The microarray is a glass slide on which specific DNA oligos are immobilized. Although microarrays have been designed for different purposes, the expression microarray consists of oligos that are complementary to each gene in a cell, with each gene spotted to a unique position on the slide. In order to compare two sets of samples, the mRNA from each sample is purified by poly (dT), followed by a reverse transcription into a complementary DNA. Each set is labeled with a different fluorescent dye, the labeled cDNAs are hybridized to the microarray, and the fluorescence of each dye is then measured. Differences in the intensity of each dye, and for each gene, are analyzed. The challenges in performing hybridization on a glass slide relate mainly to the long incubation times, the precise control of temperature and evaporation, and extensive washing requirements.

3.6.1
SELEX

The process of systematic evolution of ligands by exponential enrichment (SELEX) has been used to generate RNAs of diverse functions. The implication of evolving a biological molecule with a biological function has been of great interest. These

RNAs are generated in randomized pools, with selected sequences being separated by their desirable function. If the function is a binding interaction, the target ligands are functionalized to affinity columns and the RNAs of interest remain immobilized while the undesirable RNAs are washed off the column. The RNAs of interest are eluted from the column using high salt, and subjected to PCR intended to introduce random mutations. In this way, an RNA undergoes an accelerated evolution so that the desirable function–ligand binding–evolves into a better function of binding with high affinity and high specificity. To select for these more evolved functions, salt washes can be introduced to wash RNAs with lower binding affinities for the ligand.

3.7
Gel Filtration

Gel filtration chromatography is also referred to as size-exclusion chromatography, desalting, and buffer exchange. Unlike the other methods discussed in this chapter, gel filtration is a chromatography method only and is not an extraction. This means that the eluent fluid flows through the column, an injection is made at the top of the column, a separation occurs on the column, and the purified sample component is collected at the appropriate time. The terms "gel filtration" and "size exclusion" are descriptive in that compounds are separated based on their size.

Gel-filtration chromatography is a widely used for the purification and analysis of synthetic and biological polymers, such as proteins, polysaccharides, and nucleic acids. One advantage of the method is that the various solutions can be applied to the column without interfering with the filtration process, while preserving the biological activity of the particles to be separated.

The sample molecules in solution are separated by size as they pass through a column of crosslinked beads that are swollen with water and form a 3-D network. As the sample compounds pass through the column, the compound molecules can take either of two routes through the column, depending on the relative size of the molecule and the pore size of the beads. Those molecules that are larger than the pores will not enter the beads but simply travel straight through the column to elute first. The small molecules that are able to enter all of the pores in the beads will move in and out of the beads, and so move slowly through the column. The molecules of intermediate size, which can enter some but not all of the pores of the beads, will travel through the column slowly, but not as slowly as the smallest molecules.

The various sample compounds are collected as they are eluted from the bottom of the column, with fractions of a particular volume normally being collected. In order to determine the volume, the eluted fractions must be tested to determine which contain the sample RNA of interest and which contain the matrix compounds.

The volume of liquid in a packed bed of swollen chromatography gel beads can be represented mathematically as follows:

$$V_t = V_o + V_i + V_m \tag{3.11}$$

where V_o is the interstitial volume between the beads (also called the void volume), V_i is the volume of liquid within the swollen gel beads, V_m is the volume taken up by the resin matrix and V_t is the total volume occupied by the liquid within the beads, the interstitial liquid between the beads, and the volume of the polymer making up the bead. V_m is negligible, therefore:

$$V_t = V_o + V_i \tag{3.12}$$

The elution volume V_e, will equal V_t if the sample molecule is very small and can enter all of the pores. V_e will equal V_i, the interstitial volume if the molecules are excluded completely from the pores. In order to standardize the behavior between different columns of different sizes, it is useful to relate the elution volume, V_e, to the void volume, V_o, and total volume, V_t, where:

$$K = \frac{V_e - V_o}{V_t - V_o} \tag{3.13}$$

Thus, K subtracts out any variable influence in void volume between different columns or bead volumes of the same beaded media.

If a substance is excluded from the gel, it will elute with the void volume, $V_e = V_o$ and $K = 0$. If the substance is small, then $V_e = V_t$ and $K = 1$. For all other intermediate-sized substances K will be some decimal fraction between 0 and 1.0.

In practice, even a small sample molecule might pass straight by a pore, especially if the flow is too rapid, and therefore the flow through a gel filtration column should be slow and uniform. The pore distribution may also vary somewhat from bead to bead, and the elution curves will therefore resemble Gaussian distributions, with some experimentation being required to determine the volume of the fraction that is to be collected. As with other forms of chromatography, increasing the column length will enhance the resolution, while increasing the column diameter will increase the capacity of the column. The correct column packing is also important to maximize resolution, the main requirement being not to let air enter the column, or to allow the column packing to become dry.

References

1 Huber, C.G., Oefner, P.J. and Bonn, G.K. (1993) Rapid analysis of biopolymers on modified non-porous polystyrene–divinylbenzene particles. *Chromatographia*, **37**, 653.

2 Bonn, G., Huber, C. and Oefner, P. (1996) Nucleic acid separation on alkylated nonporous polymer beads, US Patent 5,585,236.

3. Huber, C.G., Oefner, P.J., Preuss, E. and Bonn, G.K. (1993) High-resolution liquid chromatography of DNA fragments on highly cross-linked poly(styrene-divinylbenzene) particles. *Nucleic Acids Res.*, **21**, 1061.
4. Nevejans, F. and Verzele, M. (1985) Swelling propensity (SP factor) of semi-rigid chromatographic packing materials. *J. Chromatogr.*, **350**, 145.
5. Nevejans, F. and Verzele, M. (1987) On the structure and chromatographic behavior of polystyrene phases. *J. Chromatogr.*, **406**, 325.
6. Nevejans, F. and Verzele, M. (1985) Porous polystyrene packings: characterization through pore structure determination. *Chromatographia*, **20**, 173.
7. Stuurman, H.W., Kohler, J., Jansson, S.O. and Litzen, A. (1987) Characterization of some commercial poly(styrene-divinylbenzene) copolymers for reversed-phase HPLC. *Chromatographia*, **23**, 341.
8. Rabel, F.M. (1980) Use and maintenance of microparticle high performance liquid chromatography columns. *J. Chromatogr. Sci.*, **18**, 394.
9. Gurev, I., Huang, X. and Horvath, C. (1999) Capillary columns with in situ formed monolithic packing for micro high-performance liquid chromatography and capillary electrochromatography. *J. Chromatogr.*, **855**, 272.
10. Premstaller, A., Oberacher, H. and Huber, C.G. (2000) High performance liquid chromatography-electrospray ionization mass spectrometry of single- and double-stranded nucleic acids using monolithic capillary columns. *Anal. Chem.*, **72**, 4386.
11. Unger, K.K. (1979) *Porous Silica*, Elsevier, Amsterdam.
12. Unger, K.K., Kinkel, J.N., Anspach, B. and Giesche, H. (1984) Evaluation of advanced silica packings for the separation of biopolymers by high-performance liquid chromatography, I. Design and properties of parent silicas. *J. Chromatogr.*, **296**, 3.
13. Danielson, N.D. and Kirkland, J.J. (1987) Synthesis and characterization of 2 µm wide-pore silica microspheres as column packing for the reversed-phase liquid chromatographic separation of peptides and proteins. *Anal. Chem.*, **59**, 2501.
14. Stout, R.W., Cox, G.B. and Odiorne, T.J. (1987) Surface treatment and porosity control of porous silica microspheres. *Chromatographia*, **24**, 602.
15. Barder, T.J., Wahlman, P.J., Thrall, C. and DuBios, P.D. (1997) Fast chromatography and nonporous silica. *LC-GC*, **15**, 918–26.
16. Chen, H. and Horvath, C. (1995) High-speed high-performance liquid chromatography of peptides and proteins. *J. Chromatogr.*, **705**, 3.
17. Stober, W., Fink, A. and Bohn, E. (1968) Controlled growth of monodisperse silica spheres in the micron size range. *J. Colloid Interface Sci.*, **26**, 62.
18. Engelhardt, H., Loew, H., Eberhardt, W. and Mauss, M. (1989) Polymer encapsulated stationary phases: advantages, properties and selectivities. *Chromatographia*, **27**, 535.
19. Nahum, A. and Horvath, C. (1981) Surface silanols in silica-bonded hydrocarbonaceous stationary phases. *J. Chromatogr.*, **203**, 53.
20. Schomburg, G., Deege, A., Koehler, J. and Bien-Vogelsang, U. (1983) Immobilization of stationary liquids in reversed- and normal-phase liquid chromatography. *J. Chromatogr.*, **282**, 27.
21. Schomburg, G., Koehler, J., Figge, H., Deege, A. and Bein-Vogelsang, U. (1984) Immobilization of stationary liquids of silica particles by γ-radiation. *Chromatographia*, **18**, 265.
22. Application Note AN120. Optimized purification of siRNA oligonucleotides using the WAVE® Oligo System. Available from Transgenomic Inc., Omaha, NE.
23. Horvath, C., Melander, W., Molnar, I. and Molnar, P. (1977) Enhancement of retention by ion-pair formation in liquid chromatography with nonpolar stationary phases. *Anal. Chem.*, **49**, 2295.
24. Kraak, J.C., Jonker, K.M. and Juber, J.F.K. (1977) Solvent generated ion exchange systems with anionic surfactants for rapid separation of amino acids. *J. Chromatogr.*, **142**, 671.

25 Kissinger, P.T. (1977) Comments on reverse-phase ion-pair partition chromatography. *Anal. Chem.*, **48**, 883.
26 Bidlingmeyer, B.A., Deming, S.N., Price, W.P., Sachok, J.B. and Petrusek, M. (1979) Retention mechanism for reversed phase ion-pair liquid chromatography. *J. Chromatogr.*, **186**, 419.
27 Kato, Y., Yamasadi, Y., Onaka, A., Kitamura, T., Hashimoto, T., Murotsu, T., Fukushige, S. and Matsubara, K. (1989) Separation of DNA restriction fragments by high-performance ion-exchange chromatography on a non-porous ion exchanger. *J. Chromatogr.*, **478**, 264.
28 Bloch, W. (1999) Precision and accuracy of anion-exchange separation of nucleic acids, US Patent 5,856,192.
29 Cox, R.A. (1968) The use of guanidinium chloride in the isolation of nucleic acids. *Methods Enzymol.*, **12B**, 120.
30 Chomczynski, P. and Sacchi, N. (1987) Single-step method of RNA isolation by acid guanidinium thiocyanate-phenol-chloroform extraction. *Anal. Biochem.*, **162**, 156.

4
RNA Extraction and Analysis

4.1
Transcription

Although RNA is involved in many cellular processes, it is recognized most for its central role in the regulation of genes. If the DNA is responsible for the storage of information pertinent for a cell to perform its functions, then protein is the manifestation of DNA, and it is the protein that performs the functions necessary for life. In the case of a complex eukaryote, such as a human, the genetic components of a skin cell, neuron, and muscle cell are identical. However, the way in which an organism creates so much cellular diversity depends upon the genes that are active, and those that are inactive. For a gene to become activated, steps must first be taken for that gene to become transcribed into an mRNA.

The activation of a eukaryotic gene is started by making the DNA accessible for transcription. The structure of a chromosome is both highly organized and dynamic; for example, a eukaryote cell must pack a 10 m length of DNA into a nucleus that is only 10 µm in diameter. This packaging of the long DNA molecules involves winding and packing the DNA around histone proteins to form nucleosome structures, which in turn organize themselves to form chromatin. The correct packaging of DNA not only makes the genes inaccessible for RNA polymerase to transcribe mRNA, but also allows organizational and methodical accessibility of the gene at the next steps. The histones are then rearranged or moved so as to reveal the particular gene to be expressed.

The transcription of DNA is a highly regulated process, and in eukaryotes the cell will invest considerable energy in generating the messenger RNA or pre-messenger RNA. The highly coordinated and complicated regulation of protein factors, chromatin remodeling factors and RNA polymerase enzymes are used to control the timing of transcription, with only designated genes being active for specific cellular events. Indeed, much research has been conducted over the past years in studying the patterns of gene regulation at the transcriptional level in order to identify exactly how a cell carries out its functions.

In order to study RNA at the transcriptional level, a repertoire of tools must be utilized which includes genetics, biochemistry, and systematic approaches. In general, there are two approaches for investigating transcription: (1) studies cen-

RNA Purification and Analysis: Sample Preparation, Extraction, Chromatography
Douglas T. Gjerde, Lee Hoang, and David Hornby
Copyright © 2009 WILEY-VCH Verlag GmbH & Co. KGaA, Weinheim
ISBN: 978-3-527-32116-2

tered around dissecting the very intricate details of the molecular mechanism of transcription; and (2) studies to identify those genes which are transcribed in a global sense, so as to understand which genes are required—and in what quantities—for each discrete function of a cell. Although the tools used for these two different approaches are similar, they require a number of modifications.

An understanding of the molecular mechanism of transcription requires first the development of an *in vitro* transcription system. But, to develop such as system the minimal components necessary for transcription must first be identified, and this is accomplished by using either a biochemical or genetic approach. The genetic approach follows one of two paths:

- The use of a genetic screen to determine the mutations that create or suppress a measurable phenotype. A reporter gene of which the transcriptional product is easily measured is first cloned into a vector containing a general promoter, and the clone is then transformed into a cell (e.g., yeast) which is subjected to a mutagen, such as ultraviolet light. The resultant yeast colonies are then assayed for a phenotype in which there is a defect in transcription of the reporter gene. The mutations are mapped first to the chromosome level, and eventually to the gene level. To confirm that a mutation in that gene confers the defect in transcription, a wild-type copy of the gene is transformed into the mutant cell, and asked if the phenotype is rescued. In this way, it is possible to discover unknown chromosomal positions that are involved in the transcription process.

- The introduction of mutations in a site-directed manner, to determine whether certain positions are involved in transcription. This approach is useful if there is evidence that certain proteins or domains of proteins are involved in transcription, to determine when it is possible to identify the gene that codes for that protein, to introduce point mutations or deletions of that gene, and then to assay the mutations for transcriptional defects. When the genes have been identified for their function in transcription, the protein product can be purified and tested in an *in vitro* transcription system.

Biochemistry represents another means of identifying the components involved with RNA transcription. A reporter system similar to that used for the genetic identification of cellular components can be used for an *in vitro* transcription assay. Here, the reporter gene would be a DNA molecule of known sequence that codes for a promoter and a gene. To assess quantitatively the transcription reaction, a primer extension assay is used. Here, the newly synthesized mRNA undergoes a reverse transcription reaction to generate cDNA. When reverse transcription is carried out in the presence of radioactive nucleotides, the resulting cDNAs can be resolved using polyacrylamide gel electrophoresis and visualized on film or with a phosphorimager. Positive or negative results are read as the presence or absence of a radioactive cDNA band. The transcription components are identified from a human cell line, such as HeLa cells, by adding fractionated extracts to the assay. These extracts must first be separated into nuclear and cytoplasmic fractions, by using liquid chromatography. Alternatively, the transcriptional components can be identified and purified using an immunoprecipitation assay. The minimal cellular components responsible for transcription are DNA, RNA polymerase, and ATP, and the basic assay can be modified to determine how the different promoters affect transcription, what further transcription factors enhance transcription, and how the transcription of chromatin is achieved. The minimal transcription reaction is built upon, so that the details of the molecular mechanism of transcription are understood from the bottom up.

4.1.1
RNA Catalysis

The discovery of the catalytic activity of RNA was a major breakthrough, when the catalytic RNAs – termed ribozymes – were first implicated in carrying out chemical reactions in the prebiotic world. These catalytic RNAs were first isolated from cells, notably from the nuclei of the protozoan, *Tetrahymena*, which was seen to be the location for the maturation of ribosomal RNAs.

One of the first steps of this processing involves splicing out the mature rRNAs from a long precursor RNA. The intervening sequence is a self-catalytic RNA called the Group I intron [1, 2]. These RNAs were transcribed *in vitro* and the splicing reactions carried out by incubation with nuclear extracts from *Tetrahymena*. The cell and nuclear extracts contained components that could, presumably, be fractionated into some form of active fraction. It transpired that the RNA sequence, under the proper buffer conditions, could carry out the splicing reaction.

In order to study the molecular mechanism of RNA catalysis, the minimal RNA – that is, the minimum sequence that still retains catalytic activity – must first be identified. This is accomplished by truncating the RNA and testing for activity until the minimal RNA sequence is identified. The procedure allows for an easy transcription or chemical synthesis *in vitro*, and also for an easier interpretation of the results. The transcripts can be generated in large quantities for biochemical and structural studies; the hammerhead ribozyme crystal structures were composed of RNA that contained the minimal sequence capable of catalyzing the cleavage reaction [3, 4]. The synthesis of RNAs can also be used to introduce mutations so as to identify important residues that are important both for stabilizing the fold of the ribozyme and coordinating the chemical reaction. The *in vitro* reaction represents the "cornerstone" for determining the molecular mechanism of catalysis.

4.1.2
RNA–Protein Complex Interactions

If it were to be considered that a world existed in which RNAs carried out the chemical reactions of life, then proteins would most likely have evolved so as to increase the diversity and chemical repertoire of the RNAs. Today, most cellular RNAs exist in RNA–protein complexes; indeed, even those RNAs that exist independently of protein will have been processed by protein at some point in their lifetime. These complexes have good features for separations, in that they are very stable, mostly globular, and can also be rather large. In order to extract RNAs that form part of a complex, it is first advantageous to purify the complex. For this, the preferred approach is ultracentrifugation through density gradients, as this provides a separation based on molecular weight and the density of the RNA–protein complex.

When generating a density gradient, sucrose is a highly suitable molecule. The Beckman SW 28 swing-bucket rotor permits the simultaneous processing of six 39 ml gradients, which first must be generated. One gradient production method involves layering 19 ml of a lower-density sucrose on top of 19 ml of a higher-density gradient in a disposable tube, which fits into the rotor buckets. So, when generating a 10–40% gradient, the 10% sucrose is layered on top of the 40% sucrose. The tube is then capped and a gradient maker used to generate the gradient, in approximately 2 min. Alternatively, the capped tube can be gently lain on its side for 2 h, and then gently lifted upright and settled overnight. Other less-expensive gradient makers are also available (e.g., the SG gradient maker from GE Healthcare).

It is important to note that density is dependent upon temperature. Although a lower temperature is required to maintain the activity of the RNA complexes, this will result in a more viscous gradient, and consequently any variations in temperature will affect the gradient's resolving power. If the gradients are to be processed at 4 °C, they should be maintained at that temperature prior to loading the samples and starting the run.

The samples applied to gradients for separation may be complex. Hence, if the separation requires a high resolution – for example, the purification of different RNA isoforms – the sample applied to the gradient should be as pure as possible, and in this case a sucrose cushion can be employed (see Figure 4.1). The cell lysate is first clarified by two low-speed spins of 15 min at 15 000 r.p.m. (17 600 g) to pellet the cell debris. The supernatant is layered on top a sucrose cushion of density lower than the RNA–protein complex of interest, in a fixed-angle ultracentrifuge tube. An overnight spin (Beckman Ti45 rotor) will cause the high-density material to be pelleted, trapping the contaminants. The pellets may then be resuspended in a low volume (ca. 1 ml), which is applied to the top of the sucrose gradient.

The sucrose gradients are processed in a swing-bucket rotor (here, the buckets swing out parallel with the horizon while spinning in the ultracentrifuge), during which time the applied sample will travel down through the gradient and come to an equilibrium point at the appropriate density. The centrifugal (g) force applied to the gradient and the density range of the sucrose gradient used are dictated by the resolution required for the separation. A steep gradient of 10–40% is appropriate for the separation of multiple species of largely varying densities, while a distinctly shallow gradient of 15–25% is appropriate for the separation of very similarly dense species.

In order to analyze a sucrose gradient, a high-density sucrose solution (the "push sucrose") is pumped into the gradient from the bottom. A syringe needle is used to puncture the tube at the bottom of the gradient, while a syringe pump slowly adds the push so as to not disturb the gradient. The top of the gradient is attached to a UV detector, and a chart recorder records the absorbance at 260 nm. Fractions

4.1 Transcription

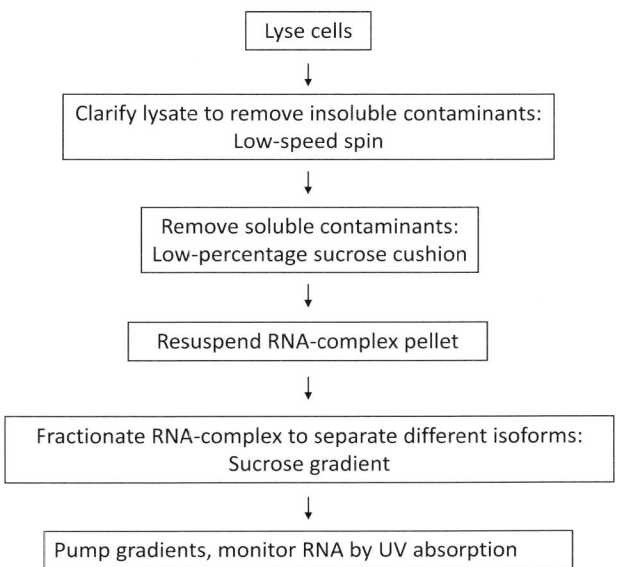

```
┌─────────────┐
│ Lyse cells  │
└─────────────┘
      ↓
┌──────────────────────────────────────────────┐
│ Clarify lysate to remove insoluble contaminants: │
│ Low-speed spin                               │
└──────────────────────────────────────────────┘
      ↓
┌──────────────────────────────────┐
│ Remove soluble contaminants:     │
│ Low-percentage sucrose cushion   │
└──────────────────────────────────┘
      ↓
┌──────────────────────────────┐
│ Resuspend RNA-complex pellet │
└──────────────────────────────┘
      ↓
┌─────────────────────────────────────────────────────┐
│ Fractionate RNA-complex to separate different isoforms: │
│ Sucrose gradient                                    │
└─────────────────────────────────────────────────────┘
      ↓
┌─────────────────────────────────────────────┐
│ Pump gradients, monitor RNA by UV absorption │
└─────────────────────────────────────────────┘
```

Figure 4.1 A schematic depiction of an ultracentrifugation method for RNA purification, using density gradients. RNAs associated in complexes can be separated from other cellular components using their high-molecular-weight properties and dense nature. Following cell lysis, the large membrane fractions are removed by a low-speed centrifugation. The clear lysates are loaded onto a sucrose cushion of moderate density. Upon ultracentrifugation, the heavy-molecular-weight components form a pellet, while other cellular components remain suspended in the sucrose cushion. The RNA complex can be resuspended and subjected to further fractionation, if required. The complexes are loaded onto sucrose density gradients which offer very high-resolution separations based on the molecular weight, density, and shape of the complexes. After ultracentrifugation, the complexes will remain suspended in the gradient, from where they can be recovered by using a gradient pump equipped with a UV detector.

can either be collected automatically, or the chart recorder can be monitored such that only the RNA–protein complexes are collected. It is possible to introduce some variables to obtain the best results; for example, the volume of the gradient can be varied depending upon the application, the sample volume, and the resolution required. Preparative-scale purifications of small (1–2 ml) samples can be processed using 38 ml gradients, whereas analytical separations, where the goal is to quantify complexes rather than to collect them, would require only small sample volumes (<1 ml) and small (19 ml) gradients to obtain sharp peaks.

4.1.3
Pre-mRNA Splicing

The product of transcription in eukaryotic cells is the pre-messenger RNA (pre-mRNA), an intermediate RNA species which must be processed before it can be translated into protein. Processing of the pre-mRNA includes adding a 5′ cap,

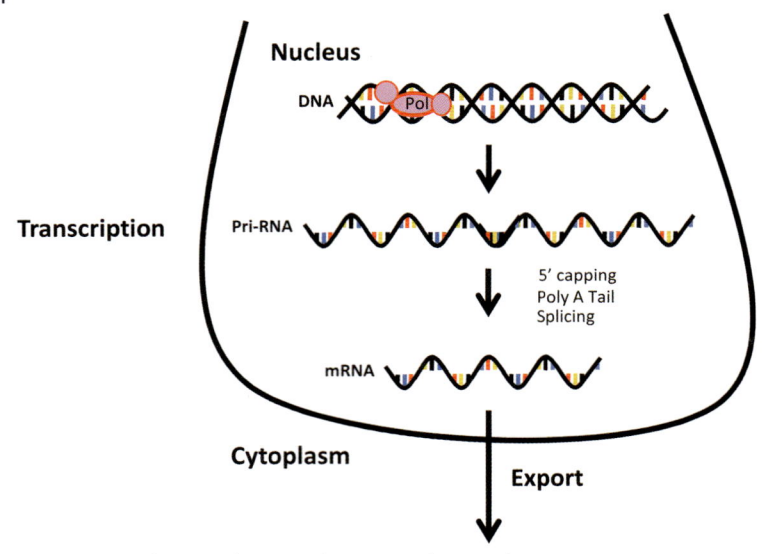

Figure 4.2 A schematic diagram of mRNA synthesis and processing depicting the steps involved in generating a mature RNA in eukaryotes. Primary mature RNA (pri-mRNA), also known as pre-messenger RNA (pre-mRNA), is first transcribed. Three sets of processing occur, including 5′ capping with a modified guanosine, 3′ polyadenylation to add a poly(A) tail, and splicing to remove introns and join the exons.

polyadenylation of the 3′ tail, and splicing out the introns (Figure 4.2). Splicing involves processing of the pre-mRNA by removing intron sequences and joining the exon sequences. These processes are performed by large macromolecular complexes called the spliceosome, which is composed of small nuclear RNAs (snRNAs) and small ribonuclear proteins (snRNPs). The structural studies of spliceosomes are complicated by the fact that these complexes are not easily purified; rather, the spliceosomes are dynamic, with different binding and unbinding events for the coordinated assembly of the complex. Strategies for the purification of spliceosomes involve manipulating the nuclear extracts to accumulate a specific complex by inhibiting specific steps of the splicing reaction. This is followed by size-exclusion chromatography and affinity tag purification. These methods have been used successfully to obtain cryo-electron microscopy structures of the spliceosome C complex [5].

4.2
Translation

When the mRNA is fully matured, its genetic information is ready for decoding by the ribosome. Much of the mechanisms of translation have been deduced from studying *Escherichia coli*. The major differences between eukaryotic and prokaryotic translation mechanisms will be described in the following text, after a general discussion of translation as understood from the *E. coli* model system.

The translation of mRNA into protein is a three-step process which is closely analogous to other polymerization reactions performed by the cell. First, the ribosome complex initiates translation; second, the mRNA is decoded and the polypeptide polymerized during elongation; and third, upon the successful completion of a protein, the translation is terminated.

During initiation in prokaryotes, the ribosome is separated into two, asymmetric nucleoprotein complexes. The large ribosomal subunit is named the 50S, and the small ribosomal subunit the 30S. The latter binds to mRNA by recognition of a specific sequence on the prokaryotic mRNA, called the Shine–Delgarno sequence. The protein initiation factors, IF-1, IF-2, and IF-3, assist by binding to the 30S ribosomal subunit; this not only prevents the 50S subunit from binding but also keeps the 30S subunit accessible for mRNA binding and increases the affinity for

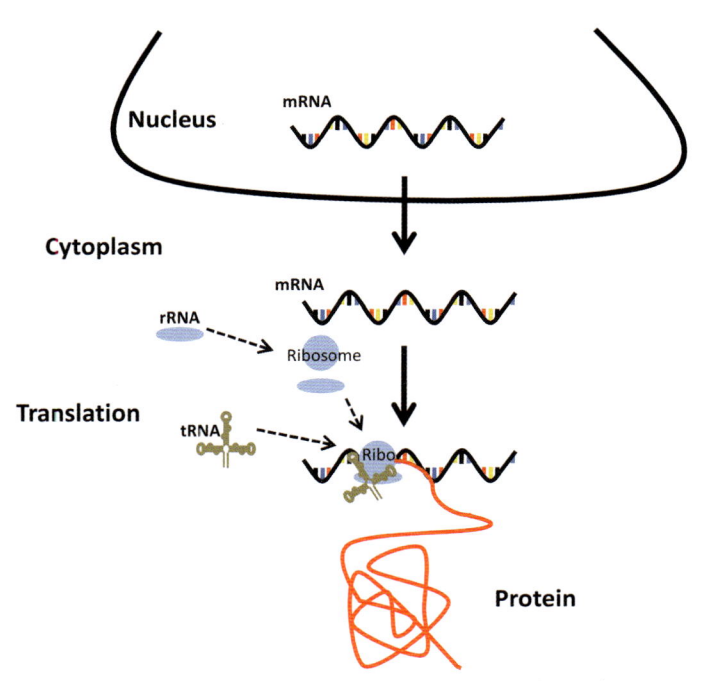

Figure 4.3 During translation of the mRNA in prokaryotes, the ribosome complex assembles onto an mRNA. Amino-acylated (aa) tRNAs bound to protein factors bring substrates for the polymerization of a polypeptide. When a proper codon–anticodon helix has formed in the ribosomal A site, the aa-tRNA is allowed to enter the ribosome, bringing the amino acid into the catalytic center of the ribosome. Peptide bond formation occurs to lengthen the polypeptide by one amino acid. The ribosome then prepares for the next polymerization event by moving the tRNAs and mRNA in a coordinated fashion so as to allow for entry of the next aa-tRNA.

the mRNA. The initiation factors also assist in recruiting the binding of the initiator tRNA, *N*-formyl-methionyl-tRNA^{fMet}, which carries the universal starting amino acid methionine that has been formylated to distinguish it from elongator methionines. Finally, protein initiation factors are released which allow the 50S ribosomal subunit to bind (Figure 4.3). The formation of the 30S ribosomal subunit, 50S ribosomal subunit, mRNA, and *N*-formyl-methionyl-tRNA^{fMet} (f-Met-tRNA^{fMet}) is known as the "70S ribosome initiator complex", and is ready for translation elongation.

During the elongation phase of translation, the ribosome polymerizes the polypeptide chain to grow a full-length, functional protein. These steps are accomplished by a precise and coordinated system of molecular interactions. The mRNA is positioned in the ribosome via the Shine–Delgarno hybridization with a sequence on the 16S rRNA of the small subunit, called the anti-Shine–Delgarno sequence. The position of that helix and the positioning of the initiator tRNA on the AUG start codon establishes the reading frame of the mRNA. This reading frame must be maintained, otherwise serious mutations will occur and result in the production of inert or possibly even dangerous protein. The interaction between the 30S subunit and the mRNA is high-fidelity in nature, and consequently the translational machinery takes much care in maintaining the reading frame. mRNA is composed of codons, which are three-nucleotide sequences that each specify precisely one amino acid, and are arranged in sequence on the mRNA, with no intervening spaces. The ribosome then essentially "reads" the codons and assembles a polypeptide sequence accordingly by binding tRNAs at the different ribosomal binding sites. The A site is the binding site for amino-acyl tRNA, the P site binds to peptidyl-tRNA, and the E site binds deacyl-tRNA. In the initiation complex, the f-Met-tRNAfMet tRNA is positioned in the P site of the ribosome, adjacent to which is the A site, where the next codon is presented. The next aminoacylated tRNA (as specified by the codon) will bind in this binding site, with peptide bond formation occurring between the formyl-methionine and the next amino acid. This results in a peptidyl-tRNA being bound to the A site and a deacylated tRNA being bound to the P site. Those tRNAs bound to the mRNA are translocated in the ribosome so that the peptidyl-tRNA is bound to the P site, the deacyl tRNA moves to the E site, and the A site is ready for binding of the next aminoacyl-tRNA, as specified by the next codon. Assisting in the elongation phase of translation are protein elongation factors. Aminoacyl tRNAs are brought to the ribosome by EF-Tu, a protein translation factor that is responsible for translation fidelity. By binding to the amino acyl end of the tRNA, EF-Tu samples the tRNA's aminoacyl end to test for proper hydrogen binding with the codon presented at the ribosomal A site. If a cognate codon–anticodon helix is formed, the ribosome triggers EF-Tu to hydrolyze a GTP, thus releasing the aminoacyl-tRNA allowing it to fully enter the ribosome. EF-G is a protein translation factor that is responsible for translocation of the tRNAs and mRNA from one binding site to another in the ribosome.

The central position that ribosomes occupy in the "Central Dogma" makes their study particularly interesting, especially as the complicated steps of translation described above highlight the complexity of the translational machinery. Many approaches have been used to tease out the details of translation, including genetics, biochemical and biophysical analysis, structure determination, and computer simulations. The *in vitro* system is the cornerstone of any biochemical analysis, and in early reconstitution studies the ribosome was separated into a protein fraction and an RNA fraction yielding inactive ribosomes. This system, in which the components can be reassembled and the activity restored [6, 7], allows for the interrogation of specific proteins and nucleotides in the translation process.

In order to prepare the 16S rRNA and protein fractions for the assembly of 30S ribosomal subunits, the ribosomal subunits are first collected as described in Chapter 2. The protein fraction is prepared by the incubation of pure 30S ribosomal subunits in buffer containing 8 M urea and 4 M LiCl on ice for 48 h – conditions which effectively precipitate the 16S RNA. Centrifugation at 10 000 r.p.m. (8000 g) for 20 min causes the 16S rRNA to be pelleted, while the protein fraction remains in the supernatant. The 16S rRNA fraction was prepared by phenol–chloroform extraction of the 30S ribosomal subunits, as described in Chapter 2.

Variations of this method allow for further fractionation of the protein components into individual proteins, and the use of recombinant proteins in order to introduce mutations and labels [8]. The 50S ribosomal subunit reconstitution is performed in a similar way, except that the 50S subunit contains two rRNA molecules, the 5S and the 23S. The reconstitution of functionally active 50S ribosomal subunits using 23S rRNA from *E. coli* is not possible, presumably due to the importance of post-transcriptional modifications of the 23S rRNA. However, 23S rRNA sequences from a number of other organisms are suitable for reconstitutions, including *Thermus aquaticus* and *Bacillus stearothermophilus* [9, 10].

Together with reconstitution studies, *in vitro* biochemistry and structural studies have generated a clearer understanding of the molecular mechanism of translation. The general framework of translation has been continually built upon so that increasing amounts of information are uncovered. Moreover, as the ribosome field continues to expand, new questions will be generated.

This is certainly the case from the information gathered from analyzing the crystal structures of the ribosome. The role of RNA was undoubtedly apparent because the catalytic center of the ribosome was composed entirely of RNA. However, the role of ribosomal proteins in translation was brought back into the spotlight when long, unstructured protein tails domains appeared to reach deep into the ribosome and to interact with tRNAs. The *in vitro* reconstitution assays outlined here would represent a clear choice for acquiring an understanding of the role of these proteins.

Occasionally, the best tool is often the least complicated. Or, simply put, why not chop off these protein tails and ask what happens? Using a method to generate precise chromosomal deletions allows *E. coli* to constitute a mutated ribosome [11], which in turn also has the distinct advantages that phenotypic effects can be measured, while large quantities of mutant ribosomes can be produced for biochemical investigations.

4.2.1
Post-Transcriptional Control of Eukaryotic Gene Expression

Although translation initiation in eukaryotes is analogous to that in prokaryotes, considerably more complexity is involved with the regulation of eukaryotic translation. Translation initiation in eukaryotes involves the action of a number of different protein factors. First, the mRNA 5′ cap is recognized, together with a number of protein factors which bind not only the mRNA but also the 40S ribosomal subunit; this leads to the mRNA being held in place, and also prevents the binding of the large ribosomal subunit.

Certain virus RNAs contain sequences that allow for translation initiation without the 5′ cap; these are named internal ribosome entry sites (IRESs). An understanding of the mechanism of initiation by IRES sequences may lead, at least potentially, to the development of viral therapeutic drugs. However, to study these mechanisms requires a pure population of the initiation components. Purification of the eukaryotic initiation complex is difficult because a number of the components are of low abundance. However, new approaches have been developed which employ affinity purification methods to isolate the ribosome complexes, whereby RNA aptamers are engineered into either the rRNA or into mRNA substrates [12, 13]. These apatamers are specific to ligands immobilized on a column, and the reversible binding allows the purification and enrichment of ribosome complexes.

4.3
Gene Regulation

Although the control of gene regulation is largely based on the decision whether to transcribe a gene into RNA, or not, the considerable information that has been gathered points to translational regulation as a means of controlling genetic expression. The mRNAs are regulated after transcription during the pre-mRNA processing steps and at the translational level. These fully mature mRNAs are either silenced or degraded in pathways separate from RNA degradation. However, instead of being signaled for degradation by deadenylation and de-capping to signal exonuclease activity, some mRNAs may be silenced or degraded in pathways which involve their identification by other small RNAs.

4.3.1
RNA Interference Pathway

The RNA interference (RNAi) pathway provides a means for the control of gene expression at the post-transcriptional level (Figure 4.4). Small interfering RNAs (siRNAs) are short, double-stranded RNAs that control gene expression by participating in post-transcriptional gene silencing. These siRNAs begin life as long, double-stranded RNAs that are subsequently cleaved into short, double-stranded RNAs by an enzyme called "dicer". One strand of the siRNA – the guide strand – is

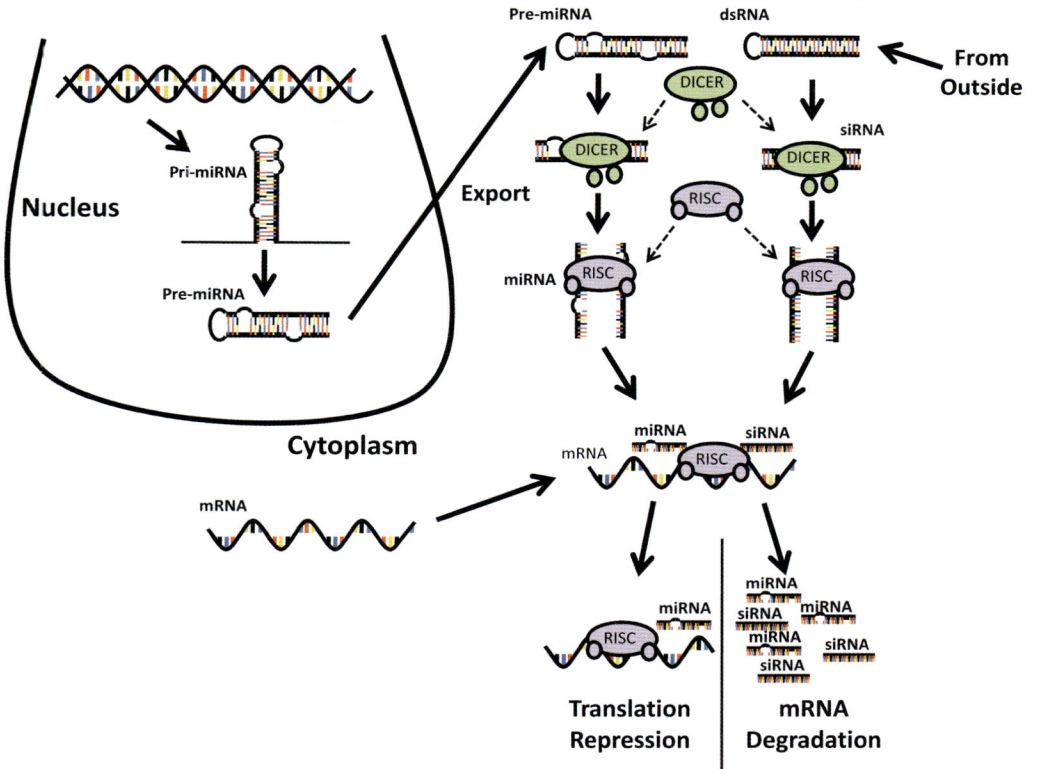

Figure 4.4 Small interfering RNAs (siRNAs) and microRNAs (miRNAs) share many components during the silencing of genes. In this schematic diagram, the differences in how the RNAs are processed are highlighted. The first step in generating an miRNA is transcription of the miRNA gene to generate a transcript called a primary microRNA (pri-miRNA). This pri-miRNA requires two processing steps. The first step excises the 3′ and 5′ ends, leaving just the stem loop, called the precursor microRNA (pre-miRNA). The pre-miRNAs are then transported out of the nucleus where other double-stranded RNAs (dsRNAs) also reside. The dicer enzyme processes both pre-miRNAs and dsRNA by generating short, single-stranded RNAs (miRNAs) and silencing RNAs (siRNAs), respectively; these associate with the RISC complex and serve to guide the ribonucleoprotein complex to RNA targets. miRNAs sometimes repress translation or signal cleavage of mRNA, while siRNAs only signal cleavage.

complementary to a sequence on the target mRNA; this sequence is incorporated into the RNA-induced silencing complex (RISC) and binds to the mRNA. The catalytic subunit, argonaute, then cleaves the mRNA, preventing the synthesis of protein.

The molecular dissection of the mechanism of RISC activity has typically been studied using *in vitro* reconstitutions of RISC, with cell extracts or components purified by immunoprecipitation. It is possible to assemble a RISC *in vitro* from completely recombinant proteins [14]. Such procedures are necessary to build a reliable *in vitro* system that allows meaningful interpretation.

4.3.2
Micro RNAs and Their Role in Gene Regulation

Micro RNAs (miRNAs) differ from siRNA in the maturation process and their mechanism of action. miRNAs are transcribed from genes, but are not translated into protein; rather, these RNAs are processed by a complex which includes the proteins Drosha (a nuclease) and Pasha (an RNA-binding protein). Once transported to the cytoplasm, the dicer processes the RNA further to form a mature miRNA. The miRNA also is incorporated in the RISC where target mRNAs are either cleaved, or translation is prevented from occurring without cleavage. Both, miRNAs and siRNAs are intriguing molecules for the possible development into therapeutic drugs.

Today, although many kits are available commercially for the purification of small RNAs, the general procedure is largely the same in all cases (Table 4.1). As

Table 4.1 Commercially available RNA purification kits.

Vendor	Name	Catalogue no.	Types of RNA targeted	General strategy
Epicentre	Masterpure yeast RNA purification kit	MPY03010	Total RNA from yeast	Salting out process for removal of proteins and other cellular contaminants. Ethanol precipitation of RNA
GE Healthcare	Illustra mRNA purification kit	27-9258-01	mRNA	Oligo(dT)-cellulose spin column purification of eukaryotic mRNAs
Invitrogen	PureLink miRNA purification kit	K1570-01	Micro and other small RNAs (<200 nt)	Phenol/guanidine (Trizol) lysis and precipitation of proteins. Large RNA depletion by silica membrane spin column (in 35% ethanol). Enrichment of small RNAs by silica membrane spin column (in 70% ethanol)
Norgen Biotek	MicroRNA purification kit	21300	Micro and other small RNAs of less than 200 nt in length	Precipitation of cell debris and contaminants for total RNA purification. Depletion of large RNAs followed by spin column purification of small RNAs
Promega	Maxwell 16 total RNA purification kit	AS1050	Total cellular RNAs	Phenol/guanidine (Trizol) lysis and precipitation of proteins. Silica membrane cartridge purification
Qiagen	miRNease Mini kit	217004	Total cellular RNAs	Phenol/guanidine lysis and precipitation of proteins. RNA enrichment by spin column purification
Qiagen	Rneasy MinElute cleanup handbook	74204	Micro and other small RNAs (<200 nt)	Further size-based fractionation of pure total RNAs

an example, the Qiagen method is as follows. The total RNA is extracted from a cellular extract in the presence of a chaotropic salt, such as guanidinium; this serves the purpose of precipitating proteins and, as a beneficial consequence, also inactivates the ribonucleases. Following phenol–chloroform phase separation, the total RNA is precipitated with an equal volume of 70% ethanol. This condition is appropriate for binding of the large RNAs to the silica membrane spin column, when the small RNAs are collected in the flow-through. To enrich the small RNA constituents, the flow-through is adjusted with a high concentration of ethanol, and applied to a second spin column where only the small RNAs are collected; this results in a fraction containing pure miRNA and siRNA.

As an alternative, HPLC represents an excellent procedure for the purification of small RNAs from complex samples (these strategies are discussed in detail in Chapter 6).

4.4
Use of siRNA to Investigate Gene Function

Aside from a potential therapeutic role, siRNA is immediately useful as a tool for investigating gene function. In the nematode *Caenorhabditis elegans*, genomic RNAi screens have been used to discover genes involved in interesting functions. For example, 16 757 *E. coli* strains, each harboring an inducible plasmid coding for an siRNA specific to each *C. elegans* gene, were used to generate a library of siRNAs. The *E. coli* were used to feed the worms, thus introducing the siRNA in a manner known to knock out the function of a specific gene. In this way, 87% of the *C. elegans* genome could be knocked out to screen for a mutant phenotype, with the genes identified presumably being important for a specific function.

References

1 Grabowski, P.J., Zaug, A.J. and Cech, T.R. (1981) The intervening sequence of the ribosomal RNA precursor is converted to a circular RNA in isolated nuclei of *Tetrahymena*. *Cell*, **23**, 467.
2 Kruger, K., Grabowski, P.J., Zaug, A.J., Sands, J., Gottschling, D.E. and Cech, T.R. (1982) Self-splicing RNA: autoexcision and autocyclization of the ribosomal RNA intervening sequence of *Tetrahymena*. *Cell*, **31**, 147.
3 Pley, H.W., Flaherty, K.M. and McKay, D.B. (1994) Three-dimensional structure of a hammerhead ribozyme. *Nature*, **372**, 68.
4 Scott, W.G., Finch, J.T. and Klug, A. (1995) The crystal structure of an all-RNA hammerhead ribozyme: a proposed mechanism for RNA catalytic cleavage. *Cell*, **81**, 991.
5 Jurica, M.S., Sousa, D., Moore, M.J. and Grigorieff, N. (2004) Three-dimensional structure of C complex spliceosomes by electron microscopy. *Nat. Struct. Mol. Biol.*, **11**, 265.
6 Traub, P. and Nomura, M. (1968) Structure and function of *Escherichia coli* ribosomes. I. Partial fractionation of the functionally active ribosomal proteins and reconstitution of artificial subribosomal particles. *J. Mol. Biol.*, **34**, 575.

7 Nomura, M. and Traub, P. (1968) Structure and function of *Escherichia coli* ribosomes. 3. Stoichiometry and rate of the reconstitution of ribosomes from subribosomal particles and split proteins. *J. Mol. Biol.*, **34**, 609.

8 Culver, G.M. and Noller, H.F. (2000) In vitro reconstitution of 30S ribosomal subunits using complete set of recombinant proteins. *Methods Enzymol.*, **318**, 446.

9 Khaitovich, P., Tenson, T., Kloss, P. and Mankin, A.S. (1999) Reconstitution of functionally active *Thermus aquaticus* large ribosomal subunits with in vitro-transcribed rRNA. *Biochemistry*, **38**, 1780.

10 Green, R. and Noller, H.F. (1999) Reconstitution of functional 50S ribosomes from in vitro transcripts of *Bacillus stearothermophilus* 23S rRNA. *Biochemistry*, **38**, 1772.

11 Hoang, L., Fredrick, K. and Noller, H.F. (2004) Creating ribosomes with an all-RNA 30S subunit P site. *Proc. Natl Acad. Sci. USA*, **101**, 12439.

12 Ali, I.K., Lancaster, L., Feinberg, J., Joseph, S. and Noller, H.F. (2006) Deletion of a conserved, central ribosomal intersubunit RNA bridge. *Mol. Cell.*, **23**, 865.

13 Locker, N. and Lukavsky, P.J. (2007) A practical approach to isolate 48S complexes: affinity purification and analyses. *Methods Enzymol.*, **429**, 83.

14 MacRae, I.J., Ma, E., Zhou, M., Robinson, C.V. and Doudna, J.A. (2008) In vitro reconstitution of the human RISC-loading complex. *Proc. Natl Acad. Sci. USA*, **105**, 512–17.

5
RNA Chromatography

5.1
Development of RNA Chromatography

The establishment of working HPLC systems for RNA Chromatography is based primarily on separation studies with single-stranded and double-stranded DNA.

Many of the early studies on liquid chromatography were conducted by J. Thompson who, together with his coworkers, published many excellent reviews [1–6] and research articles [7–9]. The reviews of 1986 and 1987 incorporated six publications, each describing a different aspect of nucleic acid separations and which, when taken together, provided a comprehensive description of the chromatographic methods available at that time for nucleic acid separations.

In 1993, Guenther Bonn, Christian Huber, and Peter Oefner first described the fundamental technology that led to modern DNA Chromatography, and later to RNA Chromatography [10–12].[1] By using ion-pairing, reverse-phase chromatography, these authors obtained rapid, high-resolution separations of both double-stranded and single-stranded DNA. The separations were performed usually in less than 10 min and, in many cases, a resolution of fragments which differed by only a single base pair in length was achieved. The rapid, high-performance separation of nucleic acids was based on the discovery and development of a C18 functionalized polymer HPLC column, later commercialized as the DNASep® column. In this procedure, an alkylammonium salt, triethyl ammonium acetate (TEAA), was added to the eluent to form neutral ion pairs when the RNA sample was introduced into the HPLC instrument (this is described later in the chapter). The fragments were adsorbed by the column, after which a gradient of water and acetonitrile was used to elute and separate the RNA fragments. As the acetonitrile

---

1) The terms, RNA Chromatography and DNA Chromatography, are commercial terms first coined by Transgenomic (Omaha, NE, USA). They are also used as academic terms. The primary reason that these terms are used rather than the technical term of reverse-phase, ion-pairing chromatography, is that it is extremely difficult (or impossible) to practice the craft of chromatography of nucleic acids without taking the precautions outlined in this book and other publications. However, once precautions regarding metal contamination, HPLC hardware, column temperature, column type and eluent are taken, RNA Chromatography becomes a reliable and powerful technology.

concentration in the eluent was increased, the smaller fragments were eluted from the column first, and in turn larger fragments were eluted and detected.

This method of ion-pairing, reverse-phase chromatography for the high-performance separation of RNA is the primary separation method described in this chapter, in Chapter 6, and also in the Appendices. Although ion-exchange chromatography can also be used to separate nucleic acid fragments with high resolution, it has the disadvantage of the fragments being difficult to collect and to purify, mainly because the eluent contains salts which must first be removed from the collected fragments before they can be used for further processing. Fortunately, the ion-pairing salts used in the studies detailed in these chapters are generally volatile, and can be evaporated relatively easily from the collected fragments.

As will be seen later in the chapter, temperature is an important HPLC parameter that can be used to control and manipulate RNA separations. Ion exchange for RNA separation is often difficult to implement at high temperatures, because high-temperature, salt-containing eluents can cause tube corrosion and plugging. Ion-pairing-type eluents can be used at high temperatures because they do not precipitate inside the HPLC tubing; neither do they have any corrosion-enhancing properties.

The new ion-pairing, reverse-phase liquid chromatography that was developed for DNA was subsequently applied to separate single-stranded and double-stranded RNAs [13, 14]. Depending on the type of eluent used and the particular RNA molecules being separated, single-stranded separations could be based on differences in the size, polarity (sequence) and shape of the molecule. Moreover, by using a larger, more nonpolar ion-pairing reagent, the RNA separation could be made to have a more size-based selectivity. However, as will be discussed later, the more complex structure of RNA compared to DNA can affect the retention times of the RNA molecules.

Recently, a number of modern commercial HPLC columns have been used for RNA separations. One such example was a silica-based C18 material (available from Varian Corporation, Walnut Creek, CA, USA), which was used in many reports of monolith polymeric materials providing excellent separations [15–17]. The performance of RNA separation columns is based on a number of properties, including the porosity of the packing material, its polarity, the absence of any metal contamination, and the small size and narrow size distribution of the packing material.

Bead pores are small relative to the nucleic acid molecule, such that any column interactions of the nucleic acid with the bead will occur on the bead surface. The polarity of the bead can be adjusted so that any nucleic acid interactions with the surface can be controlled with ion-pairing reagents. At high temperature and high pH, the silica-based beads, if not properly protected, may partially dissolve in the aqueous eluent environment. Polymer-based column materials are popular for RNA separations, not only because they are rugged but also because they can withstand extremes in eluent pH and temperature. The most popular columns for RNA separation–DNASep® and OligoSep™ (Transgenomic, Inc. Omaha, NE, USA)–have been cited in more than 1200 journal articles (www.Transgenomic.

com). Both are 2 µm, C18 surface, nonporous polymeric columns, based on the original findings of Bonn et al. [10–12].

As described in Chapter 3, the ion-pairing reagent is an amine cation salt that forms a nonpolar ion pair with the phosphate anionic group of the nucleic acid; TEAA is the most common amine cation salt used for this purpose. TEAA will pair with nucleic acid fragments to form nonpolar ion pairs which are adsorbed onto the neutral nonpolar surface of the column. Acetonitrile is then gradually added to the eluent to reduce its polarity, until the TEAA/nucleic acid ion-pair fragments are desorbed from the column. In gradient elution, the concentration of acetonitrile pumped through the column is increased gradually as the separation proceeds; in this way, the smaller fragments are eluted first, with subsequent fragments of increasing size being eluted as the acetonitrile concentration is raised.

Interestingly, the RNA fragments are extremely large molecules to be considered for separation by HPLC. The name of ribonucleic acid (RNA) is derived from the sugar group in the molecule's backbone, ribose and, depending on its biological function, the polymer will vary in length. Although RNA can exist as either double-stranded or single-stranded forms, it is normally found to be single-stranded and with a secondary structure that has a limited double-stranded character. The forms and structures of the various types of RNA found in Nature are listed in Chapter 2.

The four different nucleotide bases in RNA are adenine (A), uracil (U), cytosine (C), and guanine (G). A stylized single-stranded RNA nucleic acid molecule with an ion-paired amino alkyl reagent is illustrated below:

```
       A                  U                  C                  G
- sugar - phosphate - sugar - phosphate - sugar - phosphate - sugar -
               N                  N                  N
             alkyl              alkyl              alkyl
```

When the ion-pairing reagent is added to the RNA molecule, the combination is nonionic, due to the negatively charged phosphate and positively charged amino alkyl ion pair being formed. As the alkyl group is identical for each nucleotide, the polarity on that portion of the molecule is identical. However, the nucleic acids will vary according to the sequence of the fragment, and their polarity can vary from molecule to molecule. Fragments containing more C and G nucleotides have a higher polarity and so will elute earlier in the chromatogram than fragments containing a higher percentage of the less-polar A and U nucleotides.

The effect is even greater if the C and G nucleotides are positioned at the end(s) of the fragment. This sequence-dependent effect can be lessened for single-stranded RNA if a more nonpolar (hydrophobic) ion pairing reagent such as tetrabutylammonium bromide or N-hexylammonium acetate ion is used instead of the more commonly used TEAA. For single-stranded RNA, the separations are not solely size-based when using the TEAA eluent, because the polarity of the sequence of the fragment is not shielded by the ion-pairing reagent. Figure 5.1 shows a

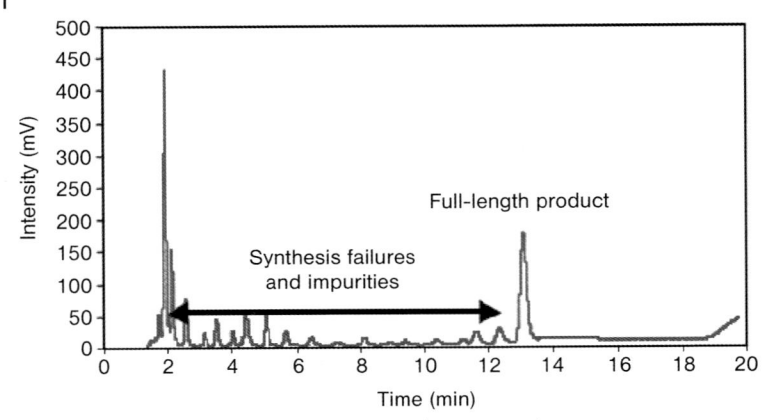

Figure 5.1 Example of chromatogram using the optimized siRNA reverse-phase denaturing HPLC purification protocol described here for the WAVE Oligo System (siRNA oligonucleotide: Luciferase Antisense) [14]. The N-hexylammonium acetate ion-pairing agent separates primarily based on length rather than on base composition. The initial product purity was <50%, but was enriched to >95%.

separation and purification of a siRNA synthesized material separated on a Transgenomic OligoSep Prep HC cartridge, using another hydrophobic reagent, n-hexylammonium acetate as the ion-pairing reagent. As the separation is primarily size-based, any synthesis failures can be removed and the product purified to more than 95% purity [14].

The polarity of the RNA molecule is also affected by the molecular structure. An ideal polynucleotide that is completely stretched out to its fullest length can interact with the surface to the greatest extent. This interaction increases the "stickiness" of the molecule for the stationary phase, and thus increases the retention. However, if the RNA molecule has a secondary structure due to the internal bonding of one part of the RNA molecule with another part, the opportunity for the molecule to interact with the stationary phase decreases, and the column retention will decrease correspondingly. Although less common, double-stranded RNA does exist in Nature. A stylized, double-stranded RNA molecule with an amino alkyl ion-pairing reagent on both strands is shown below:

```
            alkyl              alkyl              alkyl
              N                  N                  N
  - sugar - phosphate - sugar - phosphate - sugar - phosphate - sugar -
              U                  A                  G                  C


              A                  U                  C                  G
  - sugar - phosphate - sugar - phosphate - sugar - phosphate - sugar -
              N                  N                  N
            alkyl              alkyl              alkyl
```

In this case, the nucleotide bases are more internal to the molecule and are shielded, whereas the nonpolar alkyl groups are more exposed to the outside of the molecule. This has two effects:

- The alkyl group has the highest affinity for the column over the remainder of the molecule. For the same length, double-stranded nucleic acids will be retained on the column to a higher degree than will single-stranded nucleic acids.

- The double-stranded separation is size-based rather than sequence-based. However, when a secondary structure is introduced into double-stranded RNA, retention time on the column is lowered. The shape of the RNA molecule can also decrease the retention time on the column.

5.2
RNA Chromatography Instrumentation

The components of an RNA HPLC system are described in detail in Appendix 2. The components of the RNA instrument are quite similar to what is used for standard HPLC, but with some important refinements in the general flow path, oven, and detector. In many cases, the nucleic acid fragments are separated and collected for use in further research.

Contaminating metal ions (e.g., colloidal iron) may be released from the column frits, travel to other parts of the HPLC, and then become trapped. Consequently, these types of contaminant will either interfere with RNA in solution, or after it has been released and trapped on a critical component of the HPLC, including the column, an inline filter in front of the detector, or at a back-pressure device located after the detector.

Iron (III) and other metals can form insoluble complexes with phosphate anions. Hence, it is likely that surface metal ions and/or colloidal iron will combine with one or more phosphate groups on the nucleic acid fragments. If this happens, the ion-pair process chromatographic separation will be interrupted, and this will cause the peaks to broaden. In extreme cases, such metal contamination may be so severe that the nucleic acid fragments are completely prevented from eluting from the system, such that no peaks are detected.

Separations involving double-stranded RNA are more susceptible to contamination than those with single-stranded RNA. A failure to maintain the instrument and column in a clean state is probably the most common error that is made. Further details on the procedures and recipes for instrument and column maintenance are described in Appendix 3 of this book (see also [18]).

5.2.1
The Column Oven

The column oven controls the temperature of the eluent and sample entering the separation column. As the fluid entering the oven compartment is normally cooler than the oven itself, there will be a lag period before the temperature of the new fluid is raised to that of the oven. In most commercial HPLC column ovens, the fluid never reaches the set point of the oven. In RNA separation instrumentation, a pre-heating tube is used to raise the fluid temperature to that of the column before it reaches the column. The oven remains one of the most critical and difficult parameters to control in the RNA chromatograph, as the oven temperature should be accurate, precise, and not drift.

5.2.2
Ultraviolet (UV) and Fluorescence Detection

Nucleic acids absorb light strongly in the UV region, with a maximum wavelength of 260 nm. Although variable-wavelength detectors are generally set at 260 nm for detection, single-wavelength detectors function very well at 254 nm. Fluorescence detection may also be used, provided that fluorescence tags are added to the RNA. The most common dyes used in molecular biology include carboxy-fluorescein (FAM™), tetrachlorofluorescein (TET), hexachlorofluorescein (HEX™), carboxytetramethyl rhodamine (TAMRA), NED™, Pacific Blue, and many others (a list of fluorescent tags is provided in Appendix 2). Nucleic acid fragments may be detected directly at the sub-nanogram level by using UV automatic detection, and at levels which may be 10- to 100-fold lower when using fluorescence detection.

5.2.3
Fragment Collection

One of the most powerful features of RNA Chromatography is that the material can easily be purified by collecting directly from the detector effluent. Such purification can be used in biological samples where several types of nucleic acids may be present, but a particular type is desired for study. Although collection can be achieved by hand, it is normally accomplished using an automated fragment collector and controlling software. High recoveries may sometimes be limited due to a loss of the material through precipitation, or to the plating of material on the collection plate and other surfaces.

RNA is a fragile material, and may easily be degraded by enzymes following its collection. However, two factors may be of assistance for collecting stable RNA: (i) due to the purification or separation process, RNase enzymes are separated away from the fragments being collected; and (ii) the material is collected in acetonitrile, a nonpolar solvent which will denature and deactivate any enzymes that may cause degradation.

When the nucleic acid has been collected, it can be either used directly or concentrated through vacuum evaporation. Although the TEAA ion-pairing reagent is volatile and can easily be removed, other ion-pairing reagents (especially those containing quaternary ammonium groups, such as tetrabutylammonium acetate) are not volatile and cannot be removed by evaporation. In such cases the nucleic acid must be precipitated by the addition of ethanol, with even short, 20-mer single-stranded DNA being precipitated in this way.

When evaporating solutions containing TEAA, residual amounts of the ion-pairing material may be present when the vial has been evaporated to dryness. As the ionic neutrality of the nucleic acid must be maintained, each of the negative phosphate groups on the molecule must be associated with a positive counterion which, most likely, would be the triethylammonium cation.

Several reports have been made where water was used successfully to redissolve the dried material. However, if the ion-pairing reagent is present, it is likely that the recovered DNA ion-pair complex would seek out a neutral nonpolar surface once dissolved. It would seem that dissolving the nucleic acid in a buffer would provide a much more stable solution. In one procedure used successfully to recover DNA, the samples were collected from a DNA chromatogram, lyophilized on a Speedvac instrument (Thermal Savant, Holbrook, NY, USA) for 4 h at a medium heat setting, then resuspended in 50 µl of TE buffer (10 mM Tris, 1 mM disodium EDTA, pH 8.0), vortexed, and used for the appropriate application.

The collection of RNA is performed under the same constraints as DNA and several other materials. All instrumentation and solutions used for RNA must be kept free of any enzyme that will degrade RNA, and all water used with the instrument and fragment collector must be treated with diethyl pyrocarbonate (DEPC) to deactivate the RNase. Of course, the instrument should be maintained in an area which is free of dust and laboratory traffic, and gloves should be worn by the instrument operator during use in order to prevent possible RNase contamination.

Since the RNA samples cannot be evaporated due to their poor stability, they should be used directly after collection. As an example, 5 ml aliquots of collected fractions were used directly in some RT-PCR studies [19]. The precipitation of RNA was accomplished by taking an aliquot of the collected buffer, adding 10% (v/v) of a buffer (10 mM Tris–HCl, pH 7.0, 1 mM disodium EDTA, 3.0 M NaCl), followed by the addition of 1% (v/v) of an aqueous glycogen solution (10 mg ml^{-1}). After vortexing the mixture, 2.5 volumes of ethanol was added, and the solution re-vortexed. The samples were then maintained at −70 °C for 10 min, or for −20 °C for 2 h, before centrifugation at 13 000×g for 15 min at 4 °C. All precipitated RNA fractions were reconstituted in DEPC-treated water or buffer [19].

5.3
RNA Chromatography Conditions

Most of the separations reported in this chapter and in Chapter 6 were performed using the DNASep® column, 4.6 mm i.d. × 50 mm in length (Transgenomic, San Jose, USA). The stationary phase of the column consisted of a nonporous, alkylated poly(styrene-divinylbenzene) matrix, and UV detection was carried out at 260 nm. Buffer A was 0.1 M TEAA (Fluka), pH 7.0, while buffer B was 0.1 M TEAA, pH 7.0, containing 25% (v/v) acetonitrile. Note: The triethylamine used to prepare the

TEAA must be of very high quality in order to reduce any possible chromatographic background signal. It is often recommended that the eluent is purchased directly from the instrument and column manufacturers. High-quality deionized water must also be used when preparing the eluents, which must be filtered to protect the column.

A typical gradient would start at 35% buffer B, rising to 50% B in 3 min, and then to 65% B in 15 min, at a flow rate of 1.0 ml min^{-1}. The final clean-up of the column would be with 100% B for 1 min. Double-stranded RNA separations may use conditions with a slightly higher initial percentage of B, and perhaps sharper gradients. The resolution can be improved by lowering the initial percentage of B (i.e., less acetonitrile) and using shallower gradients, although the separations may take longer. Single-stranded RNA would elute from the column sooner, and under weaker eluent conditions (i.e., a lower initial acetonitrile level), and shallower gradients may be used.

It is important to note that the injection solvent may also affect the resolution. If present, the proportion of acetonitrile in the sample should be lower than the initial proportion of acetonitrile in the gradient.

Increasing the amount (and polarity) of the ion-pairing reagent in the eluent will increase the retention of all species. Sometimes, for large sample injections, the ion-pairing reagent is added to the sample to help retain the species on the column until the gradient starts.

Often, the flow rate of the eluent is lowered so as to reduce the back-pressure to the column; this is often necessary as the column ages and the back-pressure increases. The columns may be cleaned or flow-reversed in order to extend their lifetime; normally, a column would last for 1000–2000 injections, although with proper care and filtering of the samples this may be increased dramatically.

5.4
Temperature Modes of RNA Chromatography

In RNA Chromatography, the column oven controls the temperature of not only the eluent but also of the sample that is injected into the system. While RNA exists in both double-stranded and single stranded forms, and to a certain extent in both forms, it is more likely to be single-stranded but with some double-stranded secondary structure. The temperature of the column, and of the fluid entering the column, can be thought of as an additional reagent in the separation of nucleic acids, as the temperature can control whether the nucleic acid is separated as a single-molecule, as a double-stranded molecule, or as something in between.

Double-stranded nucleic acids are held together by hydrogen bonding of the two strands, but as the temperature is increased these bonds are broken, such that two single-stranded nucleic acids are created. The use of temperature is best described in the three modes of operation: nondenaturing mode; fully denaturing mode; and partially denaturing mode.

5.4.1
Nondenaturing Mode

The breaking of hydrogen bonds of double-stranded RNA through increasing the temperature is referred to as "melting". It is important that the double strand is kept intact while performing these separations, and the temperature at which melting does occur will depend on the strength of the hydrogen bonding and the environment around the RNA. Typically, a higher salt and buffer content will raise the melting temperature, whereas RNA adsorbed onto a solid surface (e.g., a column packing) will require a higher temperature before melting. In contrast, the presence of an organic solvent such as acetonitrile will lower the melting temperature, the effect being increased as the concentration of the solvent is raised. In nondenaturing mode RNA Chromatography, a high column temperature is chosen so as to lower the eluent viscosity (hence, the column back-pressure would be lowered), but not so high that denaturing would occur. The normal oven operating temperature for the nondenaturing mode in RNA Chromatography is 50 °C.

5.4.2
Partially Denaturing Mode

The partially denaturing mode is used primarily to detect mutations in DNA Chromatography through a heteroduplex detection process [20]. It should be recalled, that the hydrogen bonding which holds the two complementary strands of DNA together is specific; that is, the adenine base always bonds to thymine (and *vice versa*), while guanine always bonds to cytosine (and *vice versa*). If the two strands of a DNA fragment are perfectly matched, then they are hydrogen-bonded at each and every nucleic acid site; at this point they are said to be a homoduplex nucleic acid fragment because the two strands of the RNA are completely complementary. A heteroduplex nucleic acid fragment exists is where a genetic mutation has occurred; here, one of the bases has mutated so that a double-stranded fragment is not completely complementary but rather now contains a base that cannot hydrogen-bond to the base located on the other fragment.

Under a partially denaturing mode, differences in the melting of double-stranded DNA may be detected, because an increased melting reduces the retention time of the fragment. The operating temperature of the column and eluent must be chosen so as to partially denature or melt the fragment and enhance this mismatch. As single-stranded DNA elutes earlier than double-stranded DNA, a mixture containing both homoduplex and heteroduplex fragments can be separated, with typical oven temperatures of between 54 °C and 72 °C being used. This method is known as a "difference-detecting engine" because it detects the presence of a heteroduplex, regardless of the sequence being studied. This type of chromatography is sometimes referred to as denaturing HPLC (DHPLC) [21–23].

The partial denaturing mode in RNA Chromatography is used differently from that in DNA Chromatography. Clues to a secondary structure can be ascertained by separating a particular RNA molecule or a mixture of molecules at different temperatures. As a consequence, the retention time will be shortened as the eluent temperature is increased, indicating that a particular 3-D structure on a molecule has been changed. Fine control of the temperature may ultimately provide details of the RNA structure.

5.4.3
Fully Denaturing Mode

Both, single-stranded RNA and RNA that contains a secondary structure due to complementary sequences within its own fragment, can be also distinguished. If portions of sequences are complementary, the fragments may fold back on themselves through the formation of intramolecular hydrogen bonds, to provide several different possible structures for the same fragment. The presence of these secondary structures is both uncertain and nonreproducible, and will lead to peak shifts if they do occur. Increasing the temperature can break up the hydrogen bonding and therefore reduce some, or perhaps most, of the secondary structure [22]. However, the presence of secondary structures permits the high-resolution separation of RNA molecules. A high temperature certainly reduces sequence-specific effects, and therefore also the heterogeneity associated with such interactions with the stationary phase. However, duplex/secondary structures are present in RNA at high temperatures, and therefore play a role in maintaining good resolution. Figures 5.2a and b show the separation of an RNA ladder at 40 °C and 75 °C; here, the peaks of the RNA become sharper and more uniform as the oven temperature is increased. While it is not certain that a raised temperature will always reduce the secondary structure, in these cases the use of several different temperatures for the same sample can provide clues as to the nature of any stronger secondary structure.

5.5
Comparison of Gel Electrophoresis and Liquid Chromatography

5.5.1
Gel Electrophoresis

Gel electrophoresis is the fundamental analytical technology upon which modern molecular biology has been built. It is a staple tool in molecular biology, and critical in the success of many aspects of genetic manipulation and study.

The key aspect of gel electrophoresis is the gel material. When a potential difference voltage is applied across the electrodes, a voltage potential gradient is generated through the gel, and it is this force that drives or attracts the charged nucleic acid molecules to the electrode. The negative charges of the RNA phosphate groups are attracted to the positive electrode (anode) of the apparatus.

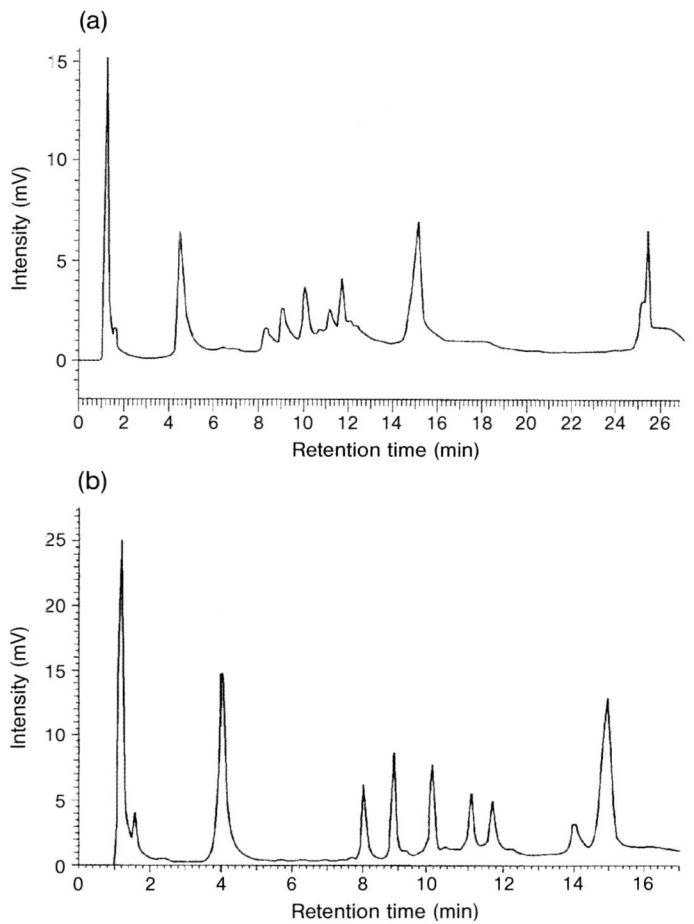

Figure 5.2 (a) Chromatogram of RNA ladder at 40 °C. RNA ladder (Cat. no. 15623010; Life Technologies) has nucleotide lengths of 155, 280, 400, 530, 780, 1280, 1520, and 1770 bases. Buffer A: 1 M TEAA, pH 7.0. Buffer B: 1 M TEAA, pH 7.0 with 25% (v/v) acetonitrile. Gradient 0.0 min 38% B, 1.0 min 40%, 16 min 60%, 22 min 66%, 22.5 min 70%, and 23 min 100%. (Reproduced with permission from [13]). (b) Same conditions as (a), except that the gradient was run at 75 °C. It is not certain that temperature will always reduce the secondary structure.

In addition, a frictional resistance slows the movement and provides the separation selectivity of the charged nucleic acids. This frictional force is a measure of the hydrodynamic size of the molecules, the shape of the molecule, and of the pore size of the medium (gel) in which the electrophoresis is taking place. As the hydrodynamic size of the molecule is closely related to the number of bases or base pairs, the separations are size-based.

The gel consists of long polymer chains that are hydrated or swollen with water; in fact, more than 95% of the gel weight is water. Two basic types of material are used to make gels, namely agarose and polyacrylamide.

- Purified agarose, a natural colloid extracted from seaweed, is generally available in powdered form and is insoluble in water or buffer at room temperature, although it will dissolve in boiling water. When the solution begins to cool, the agarose undergoes polymerization and thickens, such that the resultant polysaccharide has a molecular weight of approximately 12 000 Da and is used at concentrations of between 1% and 3% solids. Although agarose gel electrophoresis runs faster than polyacrylamide gel electrophoresis (PAGE), the resolution obtained is poorer. Agarose gels have very large effective pores and are used primarily to separate very large nucleic acid molecules of up to 20 000 base pairs, and RNA of up to several hundred base pairs.

- Polyacrylamide gels may be prepared so as to provide a wide variety of separation conditions. The pore size of the gel may be varied to produce different molecular sieving effects for separating DNAs ranging from 50 to 10 000 base pairs, and RNAs of up to several hundred base pairs, and with a better resolution than agarose. Polyacrylamide gels can be cast as either single percentage gels (i.e., no gradient), or with varying gradients. Gradient gels provide a continuous decrease in pore size from the top to the bottom of the gel, and this results in thin bands during the separation process. Polyacrylamide gels offer a greater flexibility and a more sharply defined banding than agarose gels, but are much more difficult to prepare and handle.

The quality of RNA is usually assessed using an agarose gel, the goal being to separate RNA by its size, and in most cases a traditional agarose gel is sufficient to perform such as separation. With traditional agarose gels, RNA may migrate in either a folded or semi-folded state, but this may result in smearing when the bands are visualized. In order to reduce the smearing and to visualize the banding patterns, denaturing formaldehyde agarose gel electrophoresis is commonly used. Here, the RNA is visualized under a UV light (with EtBr staining) for a quick analysis (as when using agarose to visualize DNA). Following gel electrophoresis, a Northern blot can be performed to quantify the RNA (agarose gels are not very quantitative). RNA can be extracted from the agarose gel by cutting out the band and extracting the material. In addition, the percentage crosslinking of agarose can be varied in order to optimize the separation.

Acrylamide gels can be either denaturing or nondenaturing, depending on the application. Denaturing gels are useful for size separations and offer a much better resolution than agarose gels, whereas nondenaturing gels are used for gel shift assays when there is a need to determine whether the RNA is bound to a ligand, or not. Acrylamide gels are capable of single nucleotide size separations, with a resolution capable of determining the difference between a 3′ phosphate and a 3′,5′ cyclic phosphate. It is possible to extract RNA from an acrylamide gel, but the extraction process takes longer (overnight) than extraction from an agarose gel.

Polyacrylamide gels are typically long running (a typical run may require 4 h), during which time the gel will be heated up. This is especially harmful in the case of nondenaturing gels, as it may cause the supporting glass to crack, although fortunately temperature control units are available to eliminate this problem. An additional problem is that long acrylamide gels may be difficult to handle during manipulations or transferals. For polyacrylamide gels, visualization is achieved by staining with methylene blue and using a light box (visible range). For extractions, the RNA can be visualized by its UV shadow on a fluor-coated thin-layer chromatography plate.

Gel electrophoresis can also be performed in a fused silica capillary which is filled with a hydrated linear polymer; a high voltage is then applied to the capillary's ends. Fully automated capillary electrophoresis systems have been produced to separate RNA at high resolution and high sensitivity, with fully automated sample loading. However, the separation times tend to be very long, and only small amounts of materials can be separated and purified.

5.5.2
Liquid Chromatography

RNA Chromatography can also be used to perform separations based on the size, polarity, and shape of RNA. Despite the ability of RNA Chromatography to perform size-based separations, the separation process used is entirely different compared to that of gel electrophoresis. A chromatographic separation is effected by pumping a fluid (eluent) through a column; the RNA, which is subsequently injected into the fluid stream, will stick to the column. The eluent strength is then gradually increased until the RNA fragments are eluted from the column, starting with the smallest and ending with the largest. When the separation is completed, the eluent is returned to its original strength and the whole process can be repeated.

Clearly, there are many differences between RNA Chromatography and gel electrophoresis, the major point being that the column can be thought of as a "programmable gel" in that many different types of separation can be achieved simply by choosing the appropriate set of buffer and temperature conditions. Large fragments of 50–100 base pairs can be analyzed in one run, while a few minutes later in the next run single-stranded fragment of fewer than 50 base can be measured. The conditions can be chosen for denaturing the fragment or reducing its secondary structure simply by increasing the column temperature. In addition, a gradient of elution conditions can produce any desired range of separation conditions from one run to the next. By contrast, once a gel has been made and the process of gel electrophoresis has started, no changes in the separation can be made.

One major advantage of RNA Chromatography is that it is fast–an experiment can be conceived, prepared, and analyzed within a matter of minutes–which makes the technology perfect for "what if?" types of research and development experiments. In contrast, the major benefits of gel electrophoresis are that it can be multiplexed (i.e., more than one channel can be run simultaneously), it can be used for sequencing, and large fragments can be analyzed.

Yet, RNA Chromatography is not proposed as a replacement for gel electrophoresis. The two techniques are functionally similar, as their separations are both mostly size-based, although gel electrophoresis does not result in a completely size-dependent separation of nucleic acids. This provides a significant advantage for RNA Chromatography over gel electrophoresis, in that nucleic acid separations (duplex DNA) demonstrate essentially size-dependent separations of curved or bent duplex DNA that does not migrate, as would be expected under electrophoretic conditions [24]. While these differences can translate into advantages for RNA Chromatography for many applications, both technologies have their place in the modern molecular biology laboratory, and a comparison of the two is provided in Table 5.1. However, as many of the applications discussed in this book take advan-

Table 5.1 Comparison of RNA Chromatography and gel electrophoresis.

	RNA Chromatography	**Gel electrophoresis**
Separation force	Solubility of analyte in eluent flowing through column	Electromotive attraction of analyte to electrode
Separation mechanism	Attraction of analyte to column stationary phase	Resistance of analyte to travel through gel pores
Sample separation property	Hydrophobicity or ion exchange	Hydrodynamic diameter (volume)
Primary separation effect	Size, polarity and shape, controlled with temperature and ion-pairing reagent	Size and shape, controlled with denaturing reagent
Size range	Small to medium	Small to medium
Detector sensitivity to mass amount	Excellent	Needs fluorescent tag
Effect of separation temperature	Partially and fully denaturing	Difficult to use
Gradient to control separation selectivity	Yes	Difficult
Automation	Yes	Yes with capillary, but purification still difficult
Separation speed	Fast	Medium to very slow
Multichannel	No	Yes
Purification	Yes	Difficult

tage of the unique aspects of RNA Chromatography, the reader is invited to examine their own research and to determine which technology best suits their own needs.

5.6
Analysis of Human Telomerase RNA Under Nondenaturing Conditions

A comparative example of RNA Chromatography and gel electrophoresis, as applied to RNA analysis, was presented by Mark Dickman and coworkers [25]. Human telomerase is a ribonucleprotein complex containing a catalytic protein subunit (hTERT) and an RNA species (hTR) [26, 27]. The hTR gene is transcribed by RNA polymerase II and processed at its 3′-end to produce a transcript that is 451 nucleotides in length [28–30]. The telomerase holoenzyme may act as an independent multimer (a dimer) with at least two active sites [31–33]. It has been shown that human telomerase forms an active complex containing two RNA molecules per telomerase complex [34].

The separation of hTR via nondenaturing PAGE is illustrated in Figure 5.3a, where the electrophoretogram shows the presence of two major RNA species, predicted to be the monomer and dimer of the 451 nt hTR. These results are consistent with previous findings using nondenaturing PAGE and agarose gel electrophoresis [35, 36].

An analysis of the hTR species using RNA Chromatography at temperatures between 30 °C and 75 °C gave a single peak, however, which implied that hTR is a single monomeric species under nondenaturing conditions (data not shown). The folding of RNA molecules is predominantly dependent on the presence of divalent cations. Under normal RNA chromatographic conditions, no metal ions are present in the eluent, but in this study, in order to stabilize the dimeric (or multimeric) hTR molecules under nondenaturing PAGE conditions, magnesium ions were included in the chromatography eluents. The hTR was analyzed using RNA Chromatography in the presence of $1\,mM\,Mg^{2+}$ over a range of temperatures (40–70 °C). The data in Figure 5.3b show that, at elevated temperatures of 60–70 °C, the hTR fragment elutes as a single monomeric species, but as the temperature was decreased to 40–50 °C, two peaks emerged. These results indicate that there are at least two different RNA species, consistent with those observed under nondenaturing PAGE.

A further analysis was performed to determine whether the two species separated by chromatography in the presence of magnesium were the multimeric species observed in PAGE. The two hTR species were recovered following chromatography, precipitated, and analyzed using nondenaturing PAGE. The results (Figure 5.4a and b) showed that the early-eluting RNA species (Figure 5.4a, peak "a"), when analyzed using PAGE, migrated as the monomer, whereas the later-eluting peak (Figure 5.4a, peak "b") migrated as the multimeric species (see Figure 5.4b, lane 2). These results demonstrated that, in the presence of Mg^{2+} ions, the dimer (or multimeric) hTR RNA species was stabilized and eluted after the monomer.

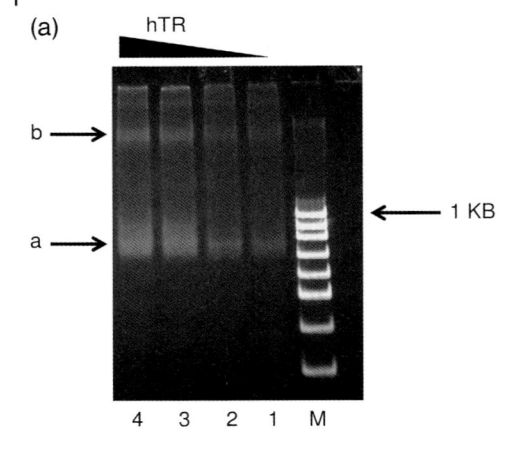

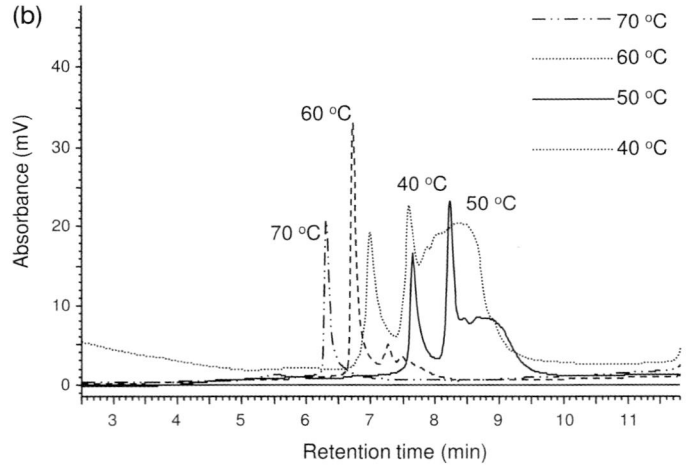

Figure 5.3 (a) Nondenaturing PAGE analysis of hTR. The electrophoretogram shows the presence of two different hTR species when analyzed under nondenaturing conditions. Lanes 1–4 contain *in vitro*-transcribed hTR, the two main RNA species are indicated by a and b. Lane M contains a 100 bp duplex DNA ladder. (Reproduced with permission from [24]). (b) RNA Chromatography separation of hTR in the presence of Mg^{2+} ions. The chromatogram shows the temperature-dependent analysis of hTR in the presence of 1 mM Mg^{2+} ions.

Following fractionation, the later-eluting RNA species (peak "b" in Figure 5.4a) was analyzed using nondenaturing PAGE. A small amount of monomer was observed, which suggested that an equilibrium existed in the multimerization of hTR. These results further supported the hypothesis that, under denaturing conditions, secondary structures of RNA transcripts remained. It appears that, by using divalent metal ions and performing the analysis under denaturing conditions, both tertiary interactions and RNA:RNA interactions are stabilized. Such interactions

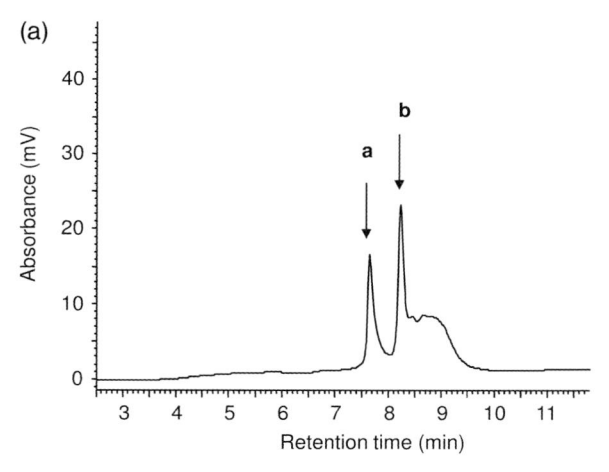

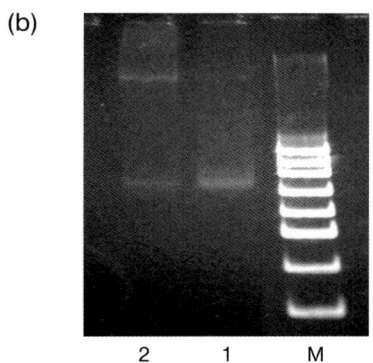

Figure 5.4 (a) The hTR species "a" and "b" were fractionated and collected using RNA Chromatography in the presence of 1 mM Mg^{2+} at 40 °C. (Reproduced with permission from Ref. [24].). (b) PAGE electrophoretogram showing the analysis of the collected hTR samples. Fraction "a" was run in lane 1, and fraction "b" in lane 2. Lane M contains a 100 bp DNA ladder.

would be possible only if the secondary structures were stable under the chromatographic conditions used in these investigations. In addition, the results of the studies showed that RNA Chromatography could be used to analyze RNA:RNA interactions as a function of temperature.

References

1 Thompson, J.A. (1986) A review of high performance liquid chromatography in nucleic acids research I. Historical perspectives. *Biochromatography*, **1**, 16.
2 Zon, G. and Thompson, J.A. (1986) A review of high performance liquid chromatography in nucleic acids research II. Isolation, purification, and analysis of oligodeoxyribonucleotides. *Biochromatography*, **1**, 22.
3 Thompson, J.A. (1986) A review of high performance liquid chromatography in nucleic acids research III. Isolation purification and analysis of supercoiled plasmid DNA. *Biochromatography*, **1**, 68.
4 Thompson, J.A. (1987) A review of high performance liquid chromatography in nucleic acids research IV. Isolation, purification, and analysis of DNA restriction fragments. *Biochromatography*, **2**, 4.
5 Thompson, J.A. (1987) A review of high performance liquid chromatography in nucleic acids research V. Nucleic acid affinity techniques in DNA and RNA research. *Biochromatography*, **2**, 68.
6 Thompson, J.A., Garfinkel, S., Cohen, R.B. and Safer, B. (1987) A review of high performance liquid chromatography in nucleic acids research VI. Nucleic acid affinity techniques in DNA-binding protein research. *Biochromatography*, **2**, 166.
7 Thompson, J.A., Blakesley, R.W., Doran, K., Hough, C.J. and Wells, R.D. (1983) Purification of nucleic acids, by RPC-5 analog chromatography: peristaltic and gravity-flow application, in *Methods in Enzymology*, Vol. 100, Academic Press, New York, pp. 368–99.
8 Green, A.P., Burzynski, J., Helveston, N.M., Prior, G.M., Wunner, W.H. and Thompson, J.A. (1995) HPLC purification of synthetic oligodeoxyribonucleotides containing base- and backbone-modified sequences. *Biotechniques*, **19**, 836.
9 Green, A.P., Prior, G.M., Helveston, N.M., Taittinger, B.E., Liu, X. and Thompson, J.A. (1997) Preparative purification of supercoiled plasmid DNA for therapeutic applications. *BioPharm*, **10**, 52–62.
10 Huber, C.G., Oefner, P.J. and Bonn, G.K. (1993) Rapid analysis of biopolymers on modified non-porous polystyrene-divinylbenzene particles. *Chromatographia*, **37**, 653.
11 Bonn, G., Huber, C. and Oefner, P. (1996) Nucleic acid separation on alkylated nonporous polymer beads. U.S. Patent 5,585,236.
12 Huber, C.G., Oefner, P.J., Preuss, E. and Bonn, G.K. (1993) High-resolution liquid chromatography of DNA fragments on highly cross-linked poly (styrene-divinylbenzene) particles. *Nucleic Acids Res.*, **21**, 1061.
13 Gjerde, D.T., Hornby, D.P., Hanna, C.P., Kuklin, A.I., Haefele, R.M. and Taylor, P.D. (2003) Method and system for RNA analysis by matched ion polynucleotide chromatography. U.S. Patent 6,576,133.
14 Application Note AN120. Optimized purification of siRNA oligonucleotides using the WAVE® Oligo System. Available from Transgenomic Inc., Omaha, NE.
15 Trojer, L., Lubbad, S.H., Bisjak, C.P., Wieder, W. and Bonn, G.K. (2007) Comparison between monolithic conventional size, microbore and capillary poly (*p*-methylstyrene-*co*-1,2-bis(p-vinylphenyl)ethane) high-performance liquid chromatography columns. Synthesis, application, long-term stability and reproducibility. *J. Chromatogr. A*, **1146** (2), 216–24.
16 Jakschitz, T.A., Huck, C.W., Lubbad, S. and Bonn, G.K. (2007) Monolithic poly[(trimethylsilyl-4-methylstyrene)-*co*-bis(4-vinylbenzyl)dimethylsilane] stationary phases for the fast separation of proteins and oligonucleotides. *J. Chromatogr. A*, **1147** (1), 53–8.
17 Wieder, W., Bisjak, C.P., Huck, C.W., Bakry, R. and Bonn, G.K. (2006) Monolithic poly(glycidyl methacrylate-co-divinylbenzene) capillary columns functionalized to strong anion exchangers

for nucleotide and oligonucleotide separation. *J. Sep. Sci.*, **16**, 2478–84.
18 Gjerde, D.T., Hanna, C.P. and Hornby, D. (2002) *DNA Chromatography*, Wiley-VCH, Weinheim.
19 Azarani, A. and Hecker, K.H. (2001) RNA analysis by ion-pair reversed phase high performance liquid chromatography. *Nucleic Acids Res.*, **29**, e7.
20 Oefner, P.J. and Underhill, P.A. (1995) Detection of nucleic acid heteroduplex molecules by denaturing high-performance liquid chromatography and methods for comparative sequencing. U.S. Patent 5,795,976.
21 Oefner, P.J. and Underhill, P.A. (1998) DNA Mutation detection using denaturing high-performance liquid chromatography, in *Current Protocols in Human Genetics* (eds N.C. Dracopoli, J.L. Haines, B.R. Korf, D.T. Moir, C.C. Morton, C.E. Seidman, J.G. Seidman and D.R. Smith), John Wiley & Sons, Inc., New York, pp. 7.10.1–7.10.12.
22 Underhill, P.A., Jin, L., Lin, A.A., Mehdi, S.Q., Jenkins, T., Vollrath, D., Davis, R.W., Cavalli-Sforza, L.L. and Oefner, P.J. (1997) Detection of numerous Y chromosome biallelic polymorphisms by denaturing high performance liquid chromatography. *Genome Res.*, **7**, 996.
23 Underhill, P.A., Jin, L., Zemans, R., Oefner, P.J. and Cavalli-Sforza, L.L. (1996) A pre-columbian Y chromosome-specific transition and its implications for human evolutionary history. *Proc. Natl Acad. Sci. USA*, **93**, 196.
24 Dickman, M.J. (2005) Effects of sequence and structure in the separation of nucleic acids using ion pair reverse phase liquid chromatography. *J. Chromatogr. A*, **1076**, 83.
25 Waghmere, S.P., Pousinis, P., Hornby, D.P. and Dickman, M.J. (2009) Studying the mechanism of RNA separations using ion pair reverse phase liquid chromatography and its application in the analysis of ribosomal RNA and RNA:RNA interaction. *J. Chromatogr. A*, **1216** (9), 1377.
26 Nakamura, T.M., Morin, G.B., Chapman, K.B., Weinrich, S.L., Andrews, W.H., Lingner, J., Harley, C.B. and Cech, T.R. (1997) Telomerase catalytic subunit homologs from fission yeast and human. *Science*, **277**, 955.
27 Beattie, T.L., Zhou, W., Robinson, M.O. and Harrington, L. (1998) Reconstitution of human telomerase activity *in vitro*. *Curr. Biol.*, **8**, 177.
28 Feng, J., Funk, W.D., Wang, S., Weinrich, S.L., Avillon, A.A., Chiu, C., Adams, R.R., Chang, E., Allsopp, R.C. and Le Yu, J.S. (1995) The RNA component of human telomerase. *Science*, **269**, 1236.
29 Zaug, A., Linger, J. and Cech, T. (1996) Method for determining RNA 3′ ends and application to human telomerase RNA. *Nucleic Acids Res.*, **24**, 532.
30 Mitchell, J.R., Cheng, J.C. and Collins, K. (1999) A box H/ACA small nucleolar RNA-like domain at the human telomerase RNA 3′ end. *Mol. Cell. Biol.*, **19**, 567.
31 Prescott, J. and Blackburn, E.H. (1997) Functionally interacting telomerase RNAs in the yeast telomerase complex. *Genes Dev.*, **11**, 2790.
32 Wenz, C., Enenkel, B., Amacker, M., Kelleher, C., Damm, K. and Lingner, J. (2001) Human telomerase contains two cooperating telomerase RNA molecules. *EMBO J.*, **20**, 3526.
33 Wang, L., Dean, S.R. and Shippen, D.E. (2002) Oligomerization of the telomerase reverse transcriptase from *Euplotes crassus*. *Nucleic Acids Res.*, **30**, 4032.
34 Gavory, G., Farrow, M. and Balasubramanian, S. (2002) Minimum length requirement of the alignment domain of human telomerase RNA to sustain catalytic activity *in vitro*. *Nucleic Acids Res.*, **30**, 4470.
35 Ly, H., Xu, L., Rivera, M.A., Parslow, T.G. and Blackburn, E.H. (2003) A role for a novel "trans-pseudoknot" RNA-RNA interaction in the functional dimerization of human telomerase. *Genes Dev.*, **17**, 1078.
36 Ren, X., Gavory, G., Li, H., Ying, L., Klenerman, D. and Balasubramanian, S. (2003) Identification of a new RNA.RNA interaction site for human telomerase RNA (hTR): structural implications for hTR accumulation and a dyskeratosis congenita point mutation. *Nucleic Acids Res.*, **31**, 6509.

6
RNA Chromatography Separation and Analysis

6.1
Features of RNA Chromatography

The retention of RNA molecules on a chromatography column is a function of the molecules' chain length, surface chemistry, and morphology. Biological RNA is found among a repertoire of intramolecular and intermolecular hydrogen-bonding interactions that give rise to structural diversity. A localized, double-stranded character is likely to be found on single-stranded RNA. One familiar structural element in an RNA molecule is the stem–loop, a noncomplementary segment of RNA which is separated by two complementary stretches of nucleotides. This, and other RNA structures, will usually shorten the molecule's retention time on the HPLC column.

In RNA Chromatography, high column temperatures can be used to denature the RNA molecule so as to control and simplify the molecular structure. However, even with thorough thermal denaturation, it is unlikely that a given population of RNA molecules can be made to be completely free from secondary and tertiary interactions. Thus, the chromatography of RNA will be different and more unpredictable than that of single-stranded and double-stranded DNA. Nonetheless, RNA Chromatography is best performed at elevated temperatures in order to denature as much as possible of the RNA being separated, and also to reduce its secondary structure.

Another useful aspect of RNA Chromatography is that it is performed using an acetonitrile solvent. The effect of acetonitrile on the activity of a typical ribonuclease (RNase) enzyme was investigated over a range of solvent concentrations and temperatures [1]. At 20% acetonitrile and 60 °C, bovine RNase was found to unfold and denature and to become an inactivate enzyme, although the activity returned (at least partially) when the temperature was reduced. However, when bovine RNase was treated with 60% acetonitrile at 60 °C it unfolded irreversibly, rendering the enzyme completely inactive. This suggests that RNA can be stabilized, or the deleterious effects of RNases can be eliminated, by the application of high temperatures and acetonitrile solvents used in RNA Chromatography.

In addition to the role of acetonitrile and temperature in RNase inactivation, the chromatographic process itself can separate RNase from the RNA fragments. For

RNA Purification and Analysis: Sample Preparation, Extraction, Chromatography
Douglas T. Gjerde, Lee Hoang, and David Hornby
Copyright © 2009 WILEY-VCH Verlag GmbH & Co. KGaA, Weinheim
ISBN: 978-3-527-32116-2

example, in a separation of RNA fragments, bovine RNase will typically elute early in the chromatogram, whereas the RNA fragments are likely to be collected later in the separation and to be completely free from the RNase molecules in the original RNA sample. RNA collected via RNA Chromatography separations has been found to be stable at room temperature, provided that the material is collected under RNase-free conditions.

6.2
Separation of Double-Stranded and Single-Stranded RNA

It has been established on many occasions that double-stranded DNA separations by DNA Chromatography are well-defined and predictable. High-resolution, size-based separations can be carried out rapidly and efficiently up to 1500 base pairs, often with signal base pair resolution for the smaller fragments [2]. The results of recent studies conducted by Mark Dickman and coworkers have shown that double-stranded RNA can also be separated over a large size range [3]. As an example, Figure 6.1 shows a 50 °C separation of double-stranded RNA and DNA, performed under identical conditions. The doubled-stranded RNA is a marker ranging from 30 to 500 base pairs, while the double-stranded DNA is a pUC18 plasmid digested with *Hae*III enzyme to produce fragments of 80 to 587 base pairs. The figure shows that the RNA fragments are eluted largely on the basis of fragment size, although fragments of RNA were eluted earlier than the corresponding DNA species of the same size. One reason for this is that RNA contains uracil rather than thymine, it having been noted that identical fragments containing uracil elute at slightly earlier retention times than do fragments containing thymine. This demonstrates that sequence-specific effects can alter the retention time of double-stranded RNA [4].

The major differences in retention time observed between the double-stranded RNA and DNA fragments cannot, however, be completely accounted for by the small differences in hydrophobicity of thymine and uracil. Double-stranded DNA is known to adopt a B-DNA conformation, whereas double-stranded RNA adopts an alternative A-DNA conformation. The A-DNA conformation is essentially a shorter, squatter version of B-DNA, and this difference in structure appears to cause a reduction in the overall double-stranded RNA hydrophobicity compared to a double-stranded DNA of the same fragment length. These results are consistent with the previous analysis of noncanonical B-DNA structures. Holliday junctions that adopt nonuniform tertiary structures, when analyzed using RNA Chromatography, showed a decrease in hydrophobicity and a corresponding decrease in retention time when compared to B-DNA of the same molecular weight [4].

As noted above, single-stranded RNA is known to adopt more stable secondary/tertiary conformations than single-stranded DNA. The high resolution of single-stranded RNA fragments, when separated at high temperatures, is demonstrated in the chromatogram of a 0.1 to 1 kb size marker shown in Figure 6.2a. Here, the separation is largely – but not entirely – size-dependent.

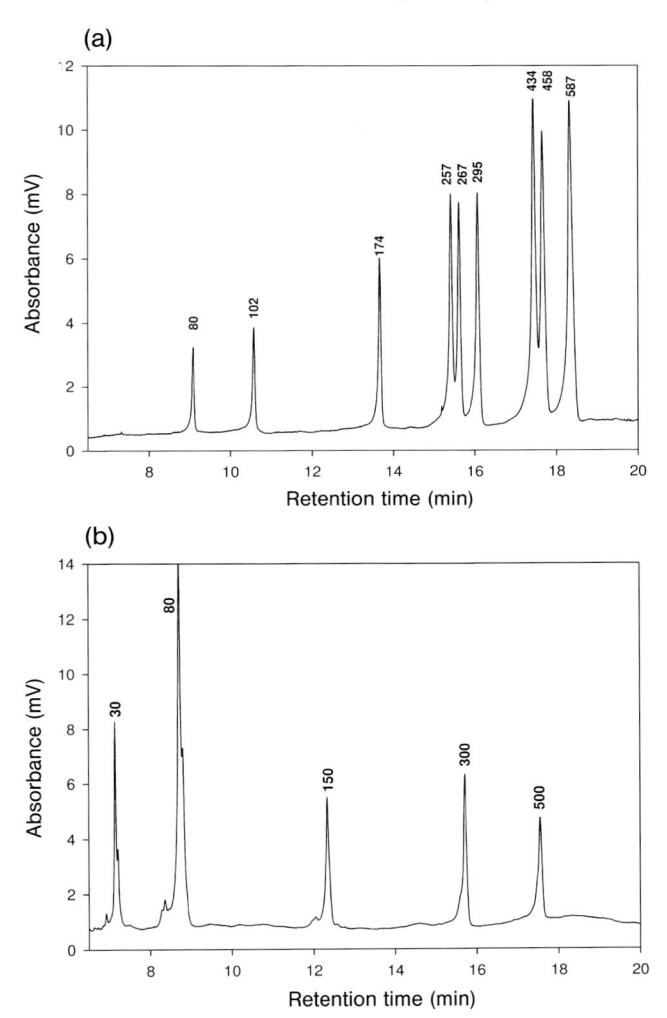

Figure 6.1 Separation at 50 °C of (a) double-stranded DNA pLC18 HaeIII digest and (b) double-stranded RNA marker. The sizes in base pairs of the nucleic acid fragments are highlighted. (Reproduced with permission from Ref. [3]).

This effect is shown more clearly in Figure 6.2b, for the analysis of total RNA extracted from mammalian cells. The separation, when performed at 75 °C, shows the coelution of 18S and 28S rRNA (1869 nt and 5035 nt, respectively), with the peaks of the high-molecular-weight RNA fragments being surprisingly sharp. It had been thought that single-stranded RNA would behave in a similar manner to single-stranded DNA, where a decrease in resolution is observed as the molecular weight of the fragments increases. This loss of resolution in the analysis of single-stranded DNA fragments is clearly observed in the analysis of a double-stranded

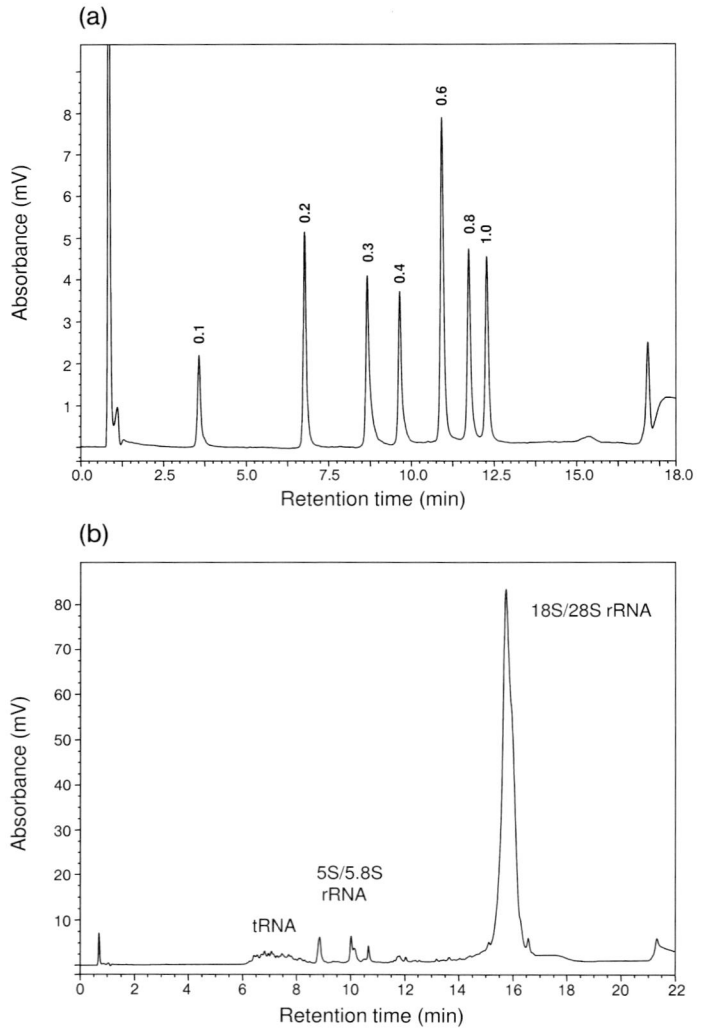

Figure 6.2 Separation of (a) single-stranded RNA and (b) total RNA. The size of the nucleic acid fragments in the RNA marker and the total RNA species are highlighted. (Reproduced with permission from Ref. [3]).

DNA pUC18 *Hae*III digest at high-temperature denaturing conditions (Figure 6.3a), where the chromatogram shows the loss of resolution for the individual DNA fragments. In contrast, the analysis of the double-stranded RNA transcripts under denaturing conditions is shown in Figure 6.3b. In this case, the results demonstrate the loss of resolution of single-stranded DNA fragment; however, an analysis of the double-stranded RNA at high temperatures reveals that each RNA duplex is resolved into single-stranded RNA fragment peaks, enabling the resolu-

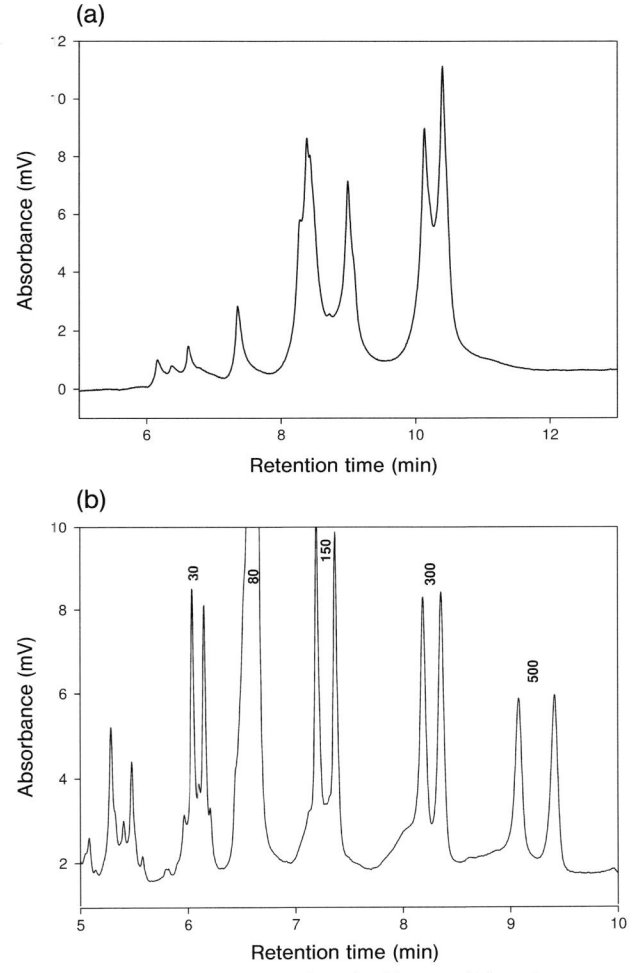

Figure 6.3 Separation at 75 °C of (a) double-stranded DNA pUC18 HaeIII digest and (b) double-stranded RNA marker. The sizes in nucleotides of the nucleic acid fragments are highlighted. (Reproduced with permission from Ref. [3]).

tion of each separate species within each duplex. This in fact provides additional proof that, with a triethyl ammonium acetate (TEAA) eluent, these separations are not completely size-dependent, as each of the two single-strand RNA fragments are generated from the same-sized double-stranded RNA duplex.

Figure 6.4 shows another separation under denaturing conditions of single-stranded RNA size standards that are commonly used in agarose gel electrophoresis. Here, the additional peak in the chromatogram is a 451 nt RNA species that forms part of the human telomerase enzyme. The telomerase RNA elutes between the RNA size standards of 200 and 500 nt, somewhat earlier than would be

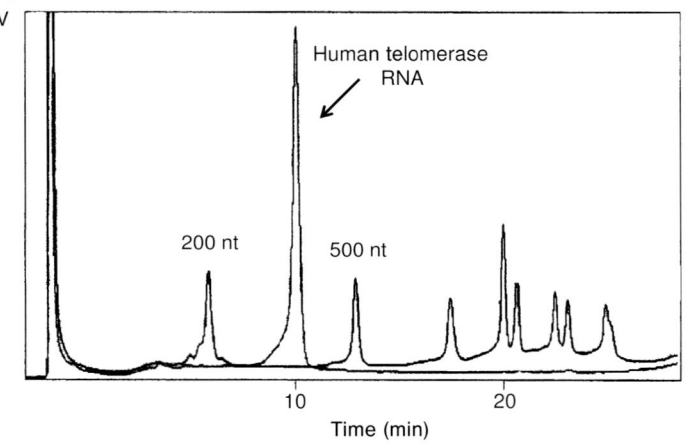

Figure 6.4 The analytical separation of RNA size standards by RNA Chromatography. The 451 nt human telomerase transcript elutes between the 200 and 500 nt species, as indicated.

expected. A DNA duplex can be considered as a negatively charged cylinder studded with a distribution of hydrophobic patches. The size-based separation of double-stranded DNA is made possible by the structural uniformity and the shielded bases of the DNA double helix. Thus, the suppression of nucleotide sequence-specific hydrophobicity is achieved by forming an ion pair with the phosphate group, with an ion-pairing reagent being used to establish the dominance of the hydrophobic (nonpolar) triethyl moiety. The hydrophobic nature of the resulting molecule is responsible for the principal analyte–stationary phase interaction of the separation.

The sequence dependence of single-stranded DNA separations can be reduced by using a more nonpolar ion-pairing reagent, such as tetrabutylammonium bromide, and increasing the hydrophobicity of the single-strand DNA/ion pair complex. It has been noted above that RNA Chromatography is best performed at elevated temperatures in order to denature and reduce the secondary structure of the RNA being separated. However, it is important to note that the electrophoresis of RNA under denaturing conditions (using urea or formamide gels) may not abolish all of the secondary and tertiary intramolecular interactions. In fact, it is impossible to predict what influence will be exerted by the temperature and solvent parameters over the chromatographic behavior of complex RNA populations. Moreover, even solutions of single molecular species should not be assumed to be conformationally homogeneous. In the language of biological NMR spectroscopy, each RNA molecule will exhibit an ensemble of conformations defined by temperature, solvent chemistry, ionic strength and, in particular, the presence or absence of divalent cations. Ultimately, any given RNA species such as a polypeptide chain will adopt its lowest free energy state(s) – a process that is thermodynamically driven by the nucleotide sequence of that particular molecule.

Therefore, when considering the mechanism of separation, the RNA structure must be taken into account. Due to the single-stranded nature of most RNA species (notwithstanding the formation of intramolecular double strands), there is an unpredictable diversity in the structure of fragments in a biological RNA sample, which makes it very difficult to elaborate any rules for RNA Chromatography. Nonetheless, much can still be learned by performing a separation under different temperatures and noting which peaks shift in terms of their retention.

6.3
Separation of Cellular RNA Species

A typical RNA Chromatography separation of a total RNA extraction is shown in Figure 6.5. There are several distinctive features within the profile (obtained in this case at 75 °C) that are found in most biological preparations of this type. The earliest eluting species include the population of tRNAs (and probably includes small nuclear RNAs), the middle section of the profile is dominated by the rRNA species, and finally underlying the entire chromatogram is a spectrum of mRNAs with many of the fragments centered on the later retention times [5]. Corroboration of the above has been obtained in the case of the tRNA fraction by comparison with commercial tRNA preparations, for mRNA by RT-PCR (see below) and for rRNA, by selectively enriching the polyadenylated mRNA population. Indeed, it is possible to determine the retention time of any RNA species (from an organism whose genome sequence is available) by RT-PCR, followed by nucleotide sequencing.

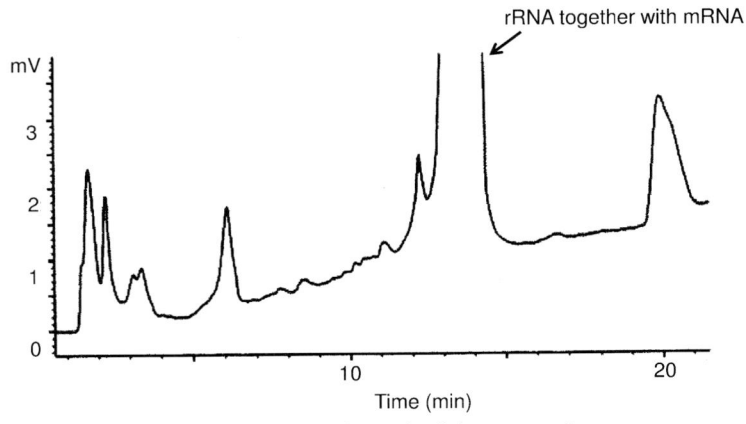

Figure 6.5 RNA Chromatography of a total cellular extract of RNA from tobacco plants. The large, broad peak eluting between 12–15 min is primarily rRNA, together with mRNA. The early-eluting peaks are tRNA, miRNA, siRNA, and other small RNAs. The large peak at 20 min is a clean up of any remaining compounds, usually large biomolecules.

Dickman and Hornby [6] described a procedure for the rapid enrichment of RNA from cell extracts. Solid-phase extraction procedures were developed utilizing nonporous alkylated poly(styrene-divinylbenzene) particles with ion-pairing reagents to enrich the total RNA. Then, by using RNA Chromatography separation and analysis, the lower-abundance small RNAs were separated from the more abundant higher-molecular-weight rRNA species.

6.4
Separation of Messenger RNA from Total and Ribosomal RNA

The expression of cellular phenotype is a direct result of the protein complement of a cell. The latter is in turn closely, but not inextricably, linked to the population of mRNA molecules produced by selective transcription of the genome in a given cell type. For this reason, considerable interest has been expressed in evaluating – in both a qualitative and a quantitative manner – the products of transcription. Although the initial experiments in this field were gene-specific, more recent methods have been introduced which facilitate population-based experiments that often utilize microarray technology.

One of the key experimental procedures that form the preludes to both of the above experiments is the systematic removal of rRNA from the mRNA (polyadenylated) fraction. Typically, the polyadenylated fraction is sequestered via an oligoT chromatography column or batch resin, the rRNA and tRNAs (together with the nonpolyadenylated mRNAs) are discarded and, after a suitable washing protocol, the mRNA fraction is concentrated and stored for subsequent experimentation. This procedure also forms the first step in the synthesis of a cDNA library (or simply cDNA) from tissue samples and for the generation of template in a typical RT-PCR experiment.

As can be seen in Figure 6.6, the outcome of a typical mRNA purification can be readily followed by RNA Chromatography. The process serves rather to enrich for mRNA than to completely remove the excess rRNA (or rather to deplete the rRNA). Indeed, at least two rounds of enrichment are usually required in order to remove the bulk of the nonpolyadenylated RNA. In this particular experiment, RNA Chromatography is used in an analytical mode for the evaluation of polyA mRNA purification. It is possible to apply the same principle in a preparative mode in order to produce a series of fractions which may facilitate the selective synthesis of cDNA populations for the production of size-selected DNA libraries.

One of the major drawbacks of polyA mRNA isolation is that some mRNAs are not polyadenylated, and therefore will be excluded from any subsequent analysis. The use of RNA Chromatography in a preparative mode, offers the potential for isolating at least a fraction of those mRNA species that do not coelute with rRNA. This is clearly an important area for development, as the analysis of cell-specific RNA populations in disease is becoming increasingly important in the field of molecular medicine.

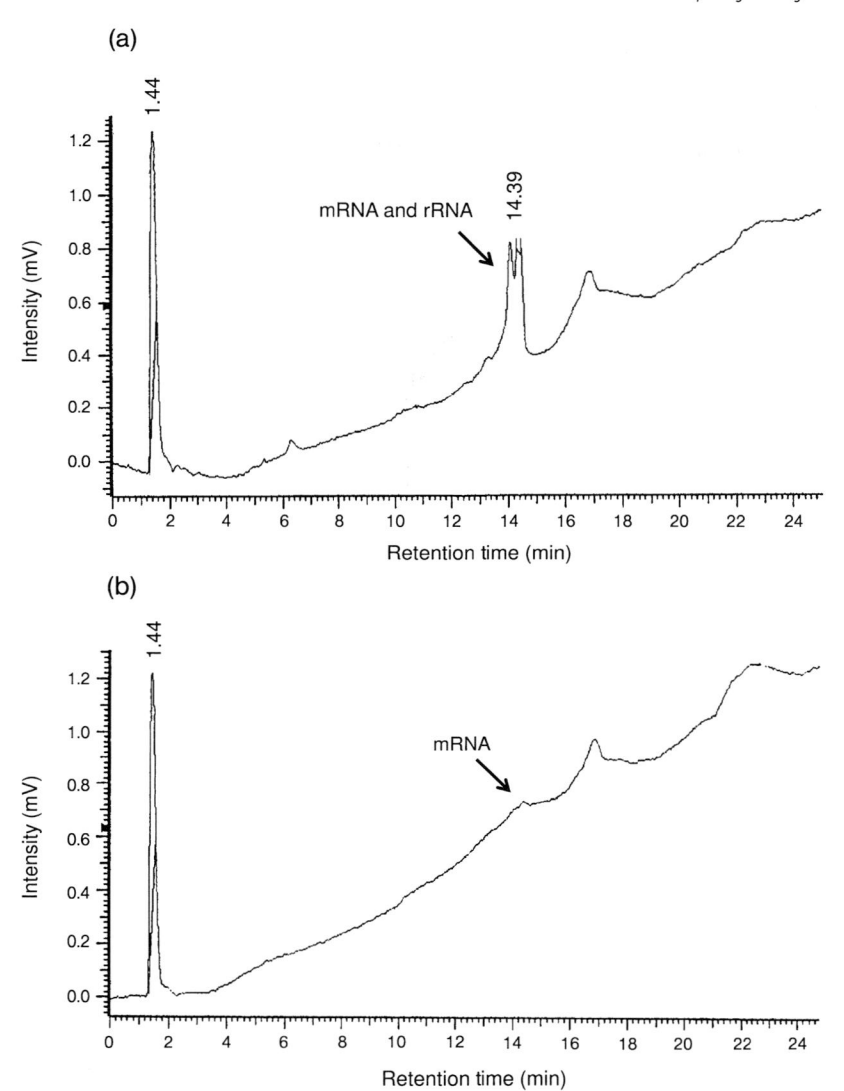

Figure 6.6 Two cycles of polyA enrichment leads to the removal of the major contaminating rRNA species. (a) The rRNA contaminant remaining after one round of enrichment on an oligodT resin; (b) rRNA contaminant remaining after a second round of enrichment.

6.5
Analysis of Transfer RNA

Each cell contains a population (usually referred to as a "pool") of tRNAs that meet the requirements of that particular cell's (or in the case of bacteria, that

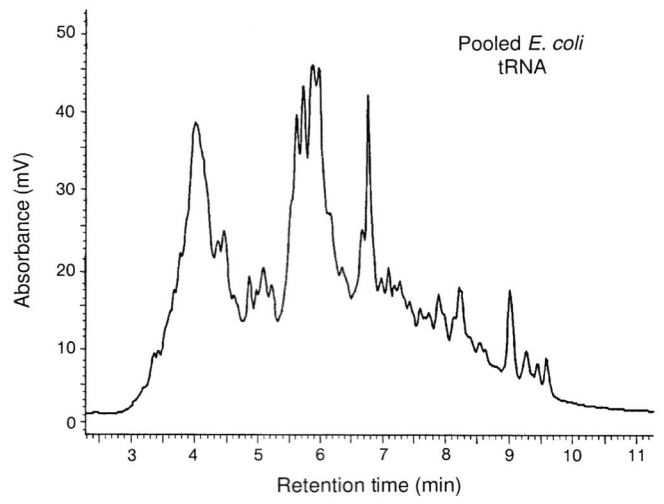

Figure 6.7 The RNA chromatogram of the entire pool of tRNAs purified from *E. coli* in denaturing. By coupling the separation with RT-PCR, it is possible to identify and quantify the condition of the individual tRNAs. (Unpublished data from M.J. Dickman and D.P. Hornby).

organism's) protein synthesis machinery. The range of sizes of tRNA molecules is particularly narrow, compared with mRNA, between approximately 60–100 nt, and a given cell typically contains about 100 species. While this molecular weight range is ideal for RNA Chromatography, many species of different conformation represent a problem for resolving the individual components in a typical cellular pool. This is clearly shown in Figure 6.7, where the total pool of tRNAs from *Escherichia coli* has been subjected to RNA Chromatography under denaturing conditions in the absence of magnesium in the sample. Nevertheless, it is clearly possible to analyze fluctuations in tRNA pools by using RNA Chromatography in conjunction with a downstream procedure such as RT-PCR.

The complexity of RNA Chromatography is exemplified by the comparison of a tRNA population under denaturing and nondenaturing conditions. The chromatography of a pooled bacterial tRNA in the presence of magnesium in the sample under denaturing conditions is shown in Figure 6.8. From this figure, it is evident that conformational factors have a dramatic influence on the chromatogram, there being much less diverse tRNA populations in Figure 6.8 than in the data shown for Figure 6.7. This difference is due to the formation and stabilization of the folded tRNA molecules, which adopt similar three-dimensional (3-D) structures. Consequently, there is a need to conduct a detailed systematic study of the temperature and solvent influences on the chromatography of complex RNA molecules, and in this respect the tRNA population represents an ideal "test bed" for the development of a robust, theoretical base for RNA Chromatography.

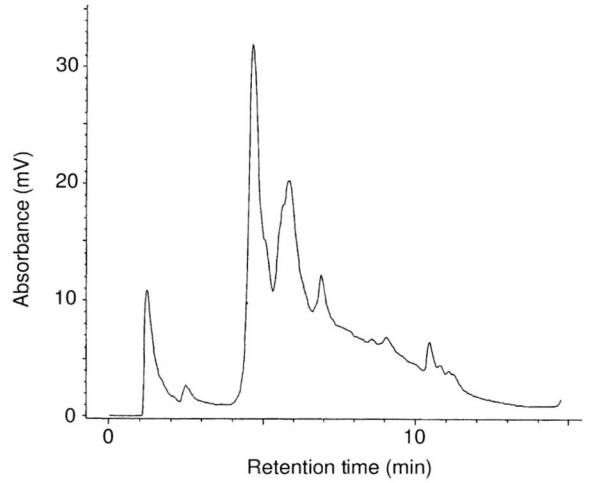

Figure 6.8 Separation of the conformational species arising from the pre-equilibration of a pooled tRNA with magnesium, under non-denaturing conditions. A much less diverse population of tRNA is observed due to the formation and stabilization of the folded tRNA molecules. The systematic variation of buffer composition and temperature provides an insight into the RNA conformation. (Unpublished data from M.J. Dickman and D.P. Hornby).

6.6
Chromatography and Analysis of Synthetic Oligoribonucleotides

One of the major contributors to the rapid rate of progress in contemporary molecular biology has been the development of automated synthetic methods for the production of oligodeoxynucleotides. These relatively short (typically between 15–100 nt in length) oligomers are used in the PCR, in numerous blotting procedures, in transcription factor biochemistry, and in some situations they have also been shown to possess catalytic activity. The development of complementary protocols for RNA synthesis has been slower to emerge, owing to problems arising through the presence of the additional 2′ hydroxyl group. Nevertheless, the routine synthesis of 20- to 30-mer oligoribonucleotides is now possible. Such oligomers, as with DNA, have great value in experimental molecular biology, but are of particular importance in the study of catalysis mediated by RNA [7]. Using the hairpin ribozyme [8] as an example, we illustrate below how RNA Chromatography can be used as a general analytical tool in studying the biochemical and conformational properties of synthetic RNA.

The susceptibility of RNA to acid–base hydrolysis provides a convenient means of optimizing the conditions for the separation of oligoribonucleotides at single-nucleotide resolution. The chromatogram shown in Figure 6.9 is of a 21 nt synthetic oligoribonucleotide which has been fluorescently labeled at its 5′ end, after exposure to 0.1 M sodium bicarbonate (pH 9 at 95 °C) for 30 min. The separation

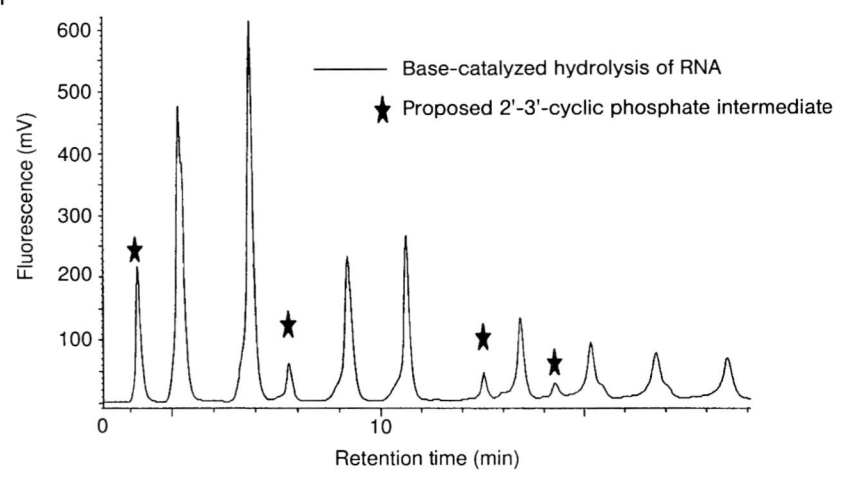

Figure 6.9 The products of an acid–base hydrolysis of a fluorescently end-labeled oligoribonucleotides. The 2′-3′ cyclic phosphate intermediate is indicated.

was developed using acetonitrile solvent and a tetrabutylammonium bromide ion pairing agent (as described in the figure legend). The high resolution of the smaller cleavage products is apparent.

The acid–base-catalyzed hydrolysis of RNA typically yields two products: a 3′ phosphate moiety; and a 5′-3′ cyclic phosphate intermediate which undergoes further nucleophilic attack to yield the 3′ phosphate product. The 3′-5′ cyclic intermediate can be visualized by using 20% denaturing polyacrylamide sequencing gels (albeit with some with difficulty). However, with RNA Chromatography, these two products may be readily identified.

The information encoded by a typical genome is initially transcribed into mRNA and subsequently translated into functional proteins (as discussed in Chapter 2). However, RNA also has the ability to catalyze biological reactions [7], this catalytic function being dependent on the 3-D shape of the RNA molecule. One technique employed to obtain RNA structural information is that of "RNA footprinting". This provides information on solvent accessibility within the RNA molecules, and can therefore be used to analyze the secondary and tertiary interactions of small RNA structures [9] and RNA–protein interactions [10]. In order to differentiate between the internal and external regions of the folded RNA molecules, the solvent accessibility of the C5′-, and also the generally quoted C4′-position of the ribose moiety, can be monitored by the addition of an Fe(II)–EDTA complex and hydrogen peroxide to the RNA in solution. The hydroxyl radicals generated primarily attack the C5′/C4′-position of the sugar, which results in cleavage of the phosphodiester bond. The cleavage products may then be analyzed directly to identify those sites that show an altered solvent accessibility.

Modifications to the standard footprinting reaction that allow the reaction products to be analyzed by RNA Chromatography, include the use of fluorescently labeled RNA. This allows the fluorescence-based detection of the cleavage products. The use of tetrabutylammonium bromide as the ion-pairing reagent is essential for the size-dependent separation of fluorescently labeled DNA, and reduces the influence of the hydrophobic fluorescent group and sequence-specific effects (see Figure 6.9). The analysis of the footprinting products is rapid, with run times of approximately 30 min for each sample. Moreover, the direct quantification of cleavage products is also possible.

In the following example, hydroxyl radical footprinting of the hairpin ribozyme (see Figure 6.10) was performed to analyze the solvent accessibility of the substrate strand as it docks in the ribozyme complex. Figure 6.11 shows the chromatogram generated by a base-catalyzed hydrolysis of the fluorescently labeled strand of the ribozyme. Separation of the cleavage products facilitates the alignment of the hydroxyl radical-generated cleavage products.

In order to analyze the solvent accessibility of the substrate strand in the folded ribozyme complex, hydroxyl radical footprinting reactions were carried out on the fluorescently labeled substrate strand. The footprinting experiment was performed in the presence of $Co(NH_3)_6^{3+}$ (cobalt hexamine is required for folding of the ribozyme into an active conformation) [12]. The results from hydroxyl radical footprinting of the fluorescently labeled substrate strand in the hairpin ribozyme complex are shown in Figure 6.12. Protection of the substrate was observed in the

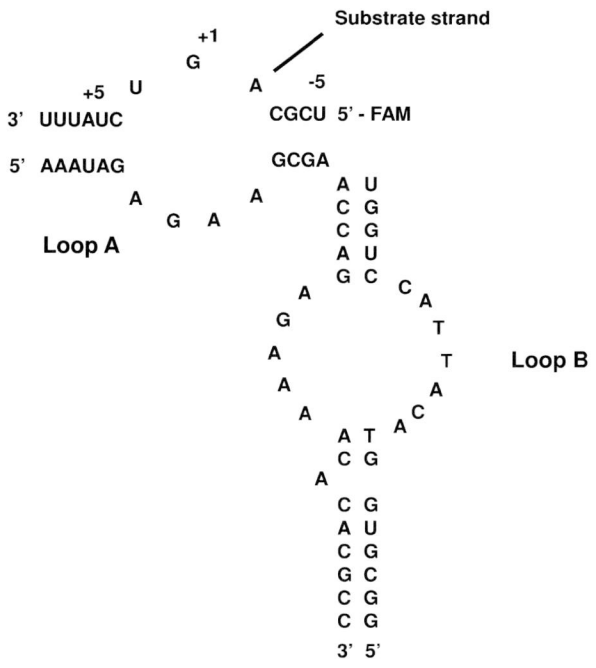

Figure 6.10 The proposed secondary structure of the hairpin ribozyme in complex with its substrate. (Reproduced with permission from Ref. [11]).

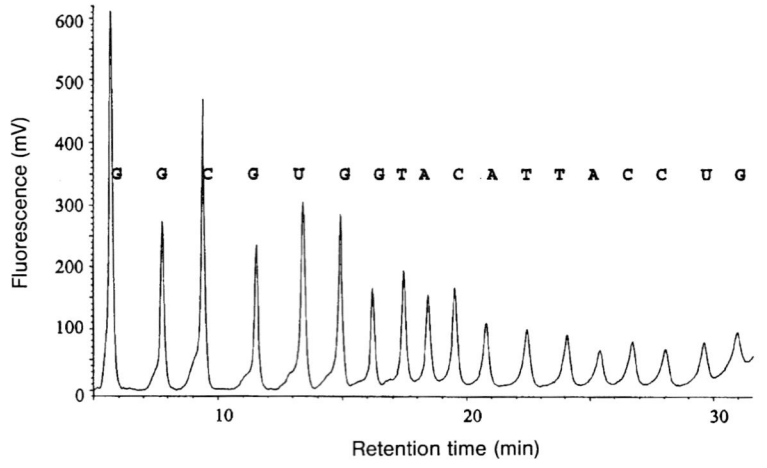

Figure 6.11 RNA Chromatography of the synthetic, fluorescent, end-labeled hairpin ribozyme following base-catalyzed hydrolysis. (Reproduced with permission from Ref. [11]).

presence of $Co(NH_3)_6^{3+}$ spanning the substrate cleavage site (a − 1, g + 1, u + 2, and c + 3). These results are fully consistent with earlier findings [12], which showed the c − 2, a − 1, g + 1, and u + 2 to be protected, thereby demonstrating that the C5′/C4′-atoms surrounding the cleavage-site ribonucleotides were internalized upon folding of the hairpin ribozyme. These results were also in complete agreement with a tertiary structure model of the hairpin ribozyme proposed in Ref. [13]. An analysis of the accessibility of the C4′/C5′-positions of the ribonucleotides in the predicted model provided a complete agreement with the experimentally observed sites of protection in the substrate strand [11]. No protection of the cleavage products was observed for the fluorescently labeled substrate strand in the absence of loop A and B RNA in the presence of Co^{2+}. Using this novel approach, RNA:RNA interactions can be analyzed in a convenient, quantitative, and high-throughput manner.

6.7
Application of RNA and DNA Chromatography in cDNA Library Synthesis

The construction of high-quality cDNA libraries is of fundamental importance in contemporary molecular biology, since such libraries play a critical role in the analysis of all aspects of gene expression. Several methods for the construction of cDNA libraries have been described [14], all of which involve a series of enzymatic reactions, including: (1) first strand synthesis, which is primed by oligo dT and catalyzed by the enzyme reverse transcriptase; (2) second strand synthesis,

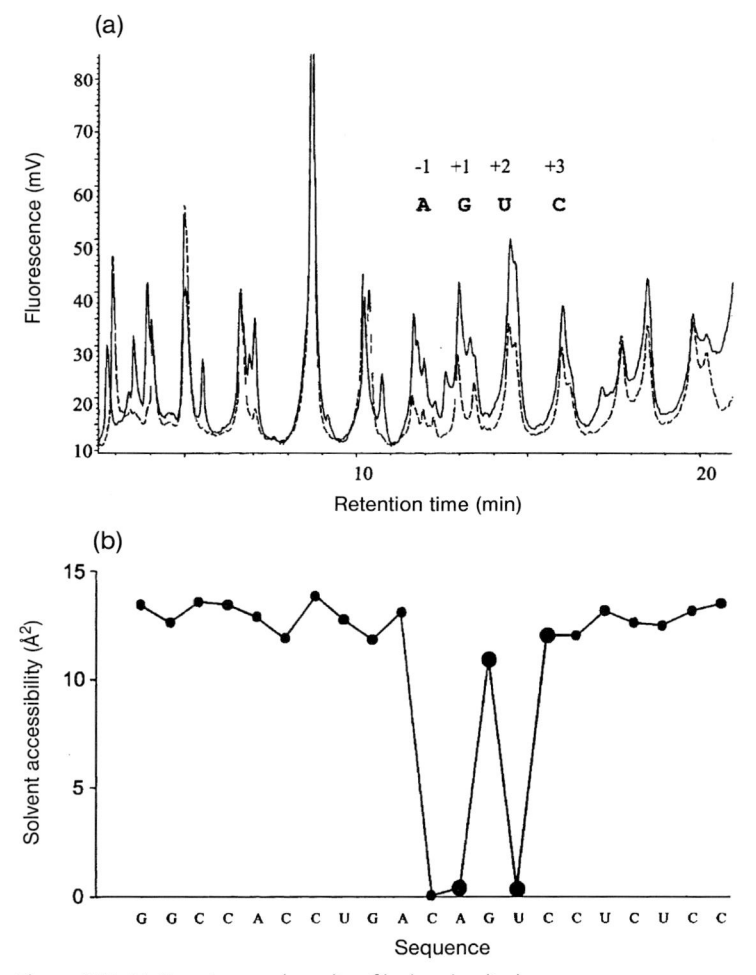

Figure 6.12 (a) Superimposed results of hydroxyl radical cleavage of the hairpin ribozyme are shown in the presence (broken line) and absence (solid line) of the substrate strand. (b) Graphical representation highlighting the theoretical solvent accessibility of bases in the complex. (Reproduced with permission from Ref. [11]).

catalyzed by a second polymerase; (3) end filling, catalyzed by Klenow DNA polymerase; and (4) DNA ligation (DNA ligase) into a vector which has often been dephosphorylated [14]. The steps in a typical protocol are shown schematically in Figure 6.13.

One major problem with this multistep procedure is the frequent failure, or suboptimal yield, of one or more of the steps. In order to ensure that all steps have been successful, careful monitoring at all stages is traditionally carried out by radiolabeling and autoradiography procedures. This normally takes one or two days to

```
Isolation of cellular RNA
          ↓
   Purification of mRNA
          ↓
  First strand cDNA synthesis
   via reverse transcriptase
          ↓                    ←──────┐
  Second strand cDNA synthesis        │
      via DNA polymerase              │
          ↓              DNA          │
                     chromatography   │
   Size fractionation by gel          │
   filtration chromatography   ←──────┘
          ↓
  Ligation into cloning vector
          ↓
 Transformation or transfection of host
          ↓
        cDNA library
```

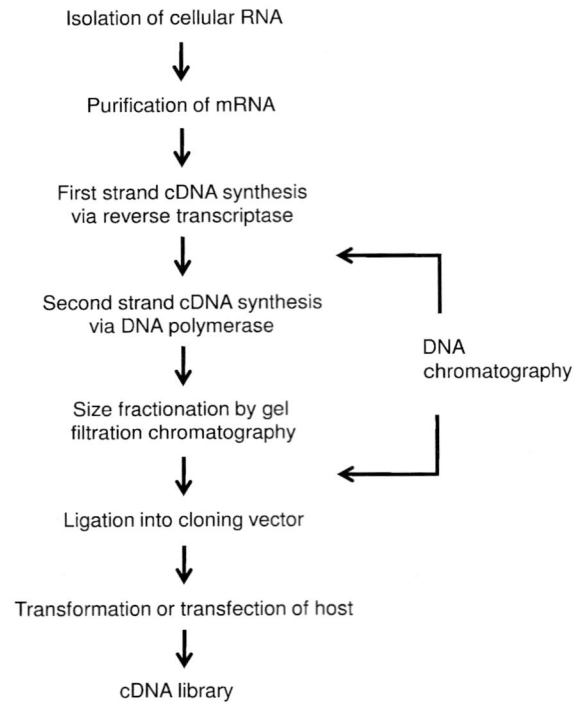

Figure 6.13 Typical steps in the preparation of cDNA. The analysis of the first strand and second strand synthetic products is greatly improved by DNA Chromatography.

complete, depending on the specific activity of the labeled material. A second problem is that when a pool of fragments is used to construct cDNA libraries the smaller fragments are selectively cloned [14]. There is a need for a size-based fractionation of the cDNA, especially when considering that the fragment size may differ in molecular weight by an order of magnitude. Sizing is traditionally achieved by gel filtration chromatography, and can be very slow. This extended use of radioactive nucleotides increases the probability of radiation damage and requires a discontinuous analysis which is usually carried out by gel electrophoresis and an ethidium bromide fluorescence assay, in which part of each fraction is lost. Clearly, there are advantages in utilizing the analytical and preparative aspects of both RNA and DNA Chromatography in order to improve these processes.

The analysis of the first strand synthesis reaction (catalyzed by reverse transcriptase) is achieved by subjecting a small fraction of the product to denaturing DNA Chromatography where, as shown in Figure 6.14, a spectrum of cDNAs should be produced. The second strand synthesis is usually accomplished through DNA polymerase I catalysis, after which Fluorogreen (Amersham) can be added (at this stage in the protocol, synthetic restriction site adapters are typically added to both ends of the double-stranded cDNA), and the cDNA fractionated by gravity flow preparative

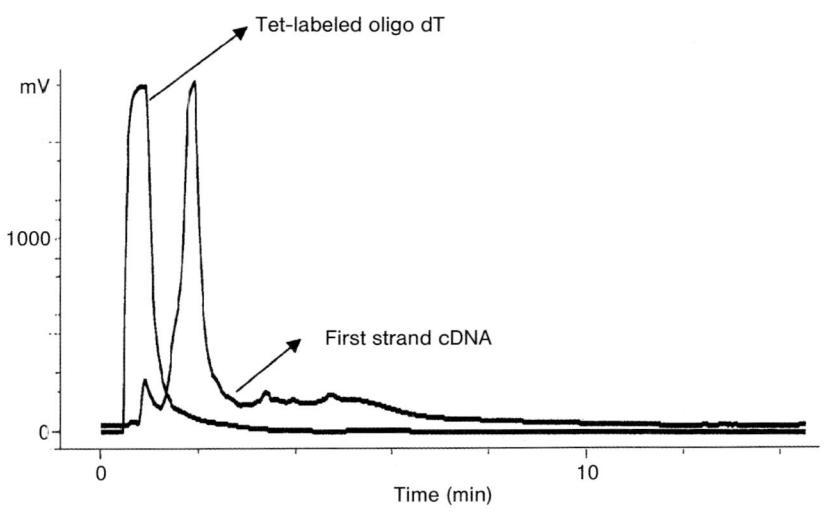

Figure 6.14 The products of a typical first strand cDNA synthesis reaction are separated using DNA Chromatography.

gel filtration. The final products of the protocol can be evaluated, both qualitatively and quantitatively, by using size-based DNA Chromatography (see Figure 6.15).

It is clearly possible, in principle, to size-fractionate the RNA immediately after isolation in a preparative procedure using RNA Chromatography. However, there are issues relating to post-column recovery that must be addressed before preparative size fractionation is used in cDNA preparation, as a total recovery of the mRNA is critical to the success of the downstream steps.

In summary, this method facilitates a rapid quality control of cDNA synthesis reactions by combining the advantages of conventional cDNA library synthesis protocols while eliminating most of their drawbacks. This approach is easy to perform, reliable, and quantitative, and also eliminates the use of radioactivity, which makes it a safe alternative to existing protocols. Moreover, there is a potential for preparative RNA fractionation prior to cDNA synthesis in order to further improve the quality of synthetic cDNA libraries enriched for "long" mRNAs.

6.8
DNA Chromatography Analyses of RT-PCR and Competitive RT-PCR Products

The value of applying DNA Chromatography to the analysis of RT-PCR products was recognized as early as 1994 [15], in a report which described the ease-of-use associated with DNA Chromatography when analyzing nested RT-PCR products (127 bp and 172 bp) derived from the 5' noncoding region of the hepatitis C virus (HCV) RNA found in human serum. The authors of the report stressed that the products could be analyzed directly by DNA Chromatography, as opposed to the requisite desalting prior to agarose gel analyses. Fluorescein-labeled primers were

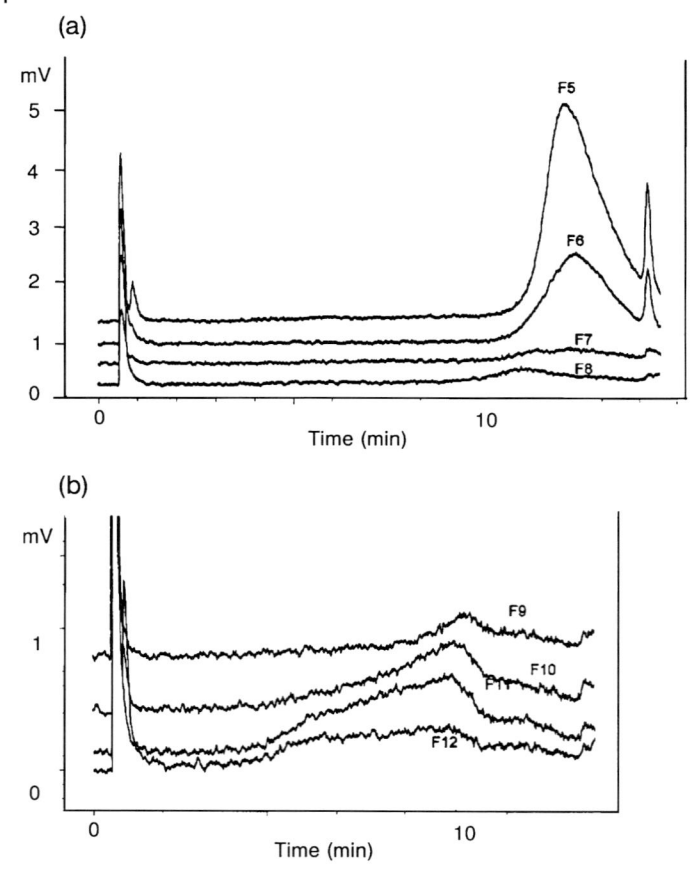

Figure 6.15 The analysis of size-fractionated products from the second strand synthesis reaction during a typical cDNA synthesis protocol.

also used, which enhanced the sensitivity 75-fold and provided a lower mass detection limit of 2 fmol. This approach not only allowed for the analysis of 90 samples in 3 h (compared to 20–40 samples in the same time period for gel-based analyses), but also provided a more accurate quantification as a result of the technique's chromatographic nature.

The results were also compared to those obtained with the enzyme-linked immunosorbent assay (ELISA; this involves the detection of an antigen – usually a macromolecule – through its recognition by a primary antibody, followed by recognition of the specific primary antibody with a "labeled" general secondary antibody). This comparison showed that not all samples containing antibodies to HCV had detectable serum-borne RNA (i.e., ELISA "false positives"), and also that antibodies had not developed in some cases of HCV RNA being detectable in the serum (i.e., ELISA "false negatives").

RT-PCR, owing to the intrinsic limitations of RNA stability in tissue samples, and the nonlinearity of PCR amplification reactions, is a difficult technique to reproduce accurately. One common means of improving the accuracy of RT-PCR gene expression quantification is to employ an exogenous, homologous "mutant" version of the native target mRNA as an internal standard. This mutant serves as competitor (internal standard) to the target mRNA in both the reverse transcription and PCR processes. The competitor is added to the sample at a known concentration. Then, assuming that the efficiency of each process is the same for the competitor and the native target mRNA, the final ratio of the peak areas of the two RT-PCR products, multiplied by the original concentration of the competitor, provides a highly accurate means for determining the original target mRNA copy number. This general process is typically referred to as "competitive RT-PCR".

A key insight regarding these analyses was provided in 1995 by Doris and coworkers [16], who observed that competitive RT-PCR products formed stable mutant:target heteroduplices. When a gel-based analysis is performed, these heteroduplices tend to comigrate with the mutant product, which leads to an error in the calculations. In the past, this source of error was dismissed as affecting mutant and target bands equally in gel-based analyses, such that the errors "cancelled each other out". However, these investigators showed that an uncorrected comigration could lead to large errors when attempting to accurately quantify the level of gene expression.

This problem was overcome by applying DNA Chromatography so that the heteroduplices were effectively and quantitatively separated from the mutant and native RT-PCR homoduplices, and the homoduplices were also resolved from one another. When the three components had been resolved and quantified in terms of peak area, it was possible to calculate the *relative* number of final copies of all three duplex fractions, as the fragment sizes and molar absorptivities were known. The heteroduplex fraction could be proportionally re-allocated to the native and mutant fractions, so that a simple target-to-competitor ratio was obtained for the purposes of quantification. This provided assays that were highly accurate (95% recovery for a known amount of starting material), very precise (coefficients of variation of 8.3% to 17.8% for analyses of rat brain and nephron tissue, respectively), linear over four orders of magnitude, and allowed so-called "single-tube" analyses (no titrations with varying concentrations of competitor were necessary). This general approach was applied in subsequent studies for the quantification of rat angiotensinogen expression [17], as well as for the quantification of endoplasmic reticulum (ER) expression in osteoblast cell lines [18]. These last two reports most likely represent the most rigorous RT-PCR analyses to date.

Clearly, the single most compelling feature of DNA Chromatography, when combined with competitive RT-PCR, is that of being able to resolve the mutant:target heteroduplices that are likely to form in many instances. However, a key aspect to this approach to gene expression analysis is that an exogenous competitor mutant mRNA sequence must be created and introduced. Furthermore, if the exogenous introduced mutant does not reverse transcribe and/or undergo the PCR with efficiencies identical to that of the target, then it must be determined exactly to what extent these efficiencies differ, and a suitable correction factor applied. This is particularly true if the aim is to perform "single-tube" assays, whereby perfect linearity is assumed for equal RT *and* PCR efficiencies for both the mutant and target. To eliminate the need for any prior characterization of the RT and PCR efficiencies, the investigators examined the effect of mutant homology [19, 20]. Their results showed that, when the mutant was a 145 bp insertion mutant (64% homology), the efficiency of reverse transcription for mutant versus target differed by 3.8-fold. However, when the mutant was a 14 bp deletion mutant, the efficiency of reverse transcription for mutant versus target was identical. Interestingly, these differences in PCR efficiency were the same for the large insertion mutant and the smaller deletion mutant. Consequently, it was determined that a high degree of homology between mutant and target would be needed to achieve a good overall accuracy.

In one of these studies [20], several of the fundamental features of DNA Chromatography-mediated RT-PCR were examined even further. Again, variable co-reverse transcription and co-amplification efficiencies between the target and competitor were investigated as a function of the insertion/deletion mutant length, in addition to various reverse transcriptase modifiers, RNA pre-heating procedures (with and without dimethylsulfoxide), and the concentrations of magnesium chloride, nucleotides, and random hexamers. While altering these conditions led to significant (albeit small) differences in RT efficiencies, the single largest contributor equalizing the efficiencies between mutant and competitor was the difference in the insertion or deletion. For example, a 24 bp insertion mutant for rat CYP, a homologue of cyclophilin (a cytosolic binding protein for cyclosporin A), provided identical RT efficiencies as compared to the native sequence; however, a 28 bp deletion mutant for rat could not be equally reverse transcribed when compared to the native sequence. The same authors also reported that DNA Chromatography with UV detection, when combined with fully optimized competitive RT-PCR procedures, would allow for the reliable detection of as few as 100 mRNA copies.

In an effort to make these analyses simple and available to investigators, a shareware Microsoft Excel Macro (Peter Doris, University of Texas at Houston) was developed for processing the integrated peak data for target, mutant, and any associated heteroduplices [21]. The user is first prompted to enter information regarding the sizes of the competitor and target and, when completed, it is possible to enter data for either titration-based analyses or single-tube analyses (peak areas, competitor mass). The mathematics analysis is performed via the Microsoft Excel macro logic, and results are reported in a spreadsheet format (along with titration plots, if so chosen). A similar program is also available from Transgenomic, Inc. (Omaha, NE, USA).

6.9
Alternative Splicing

The same authors that pioneered [15] DNA Chromatography-mediated competitive RT-PCR analyses, also reported on the DNA Chromatography-mediated determination of alternative splicing [22]. In alternative splicing, the coding sequence effectively becomes larger or smaller by changes occurring in the splicing position, thus providing numerous forms of a protein from a single gene. When expressed, this will have the effect of producing mRNA that carries an "insertion" or "deletion" of a particular length relative to the native target length, which in turn acts as an endogenous RT-PCR "competitor internal standard". This internal standard operates analogously to the exogenous competitors described above for the quantification of gene expression. As with the quantification of gene expression, accurate alternative splicing determinations require the resolution and quantification of native "competitor" heteroduplices to account for all reaction products.

The first report of this approach for detecting alternative splicing described the biological properties of the 1α and 1β isoforms of RUSH, a rabbit homologue of a human protein that binds to the human immunodeficiency virus-1 promoter [22]. These investigations required an accurate determination of the extent of one isoform's expression relative to the other, and DNA Chromatography showed RUSH 1 to be the progesterone-dependent isoform. These authors later elaborated on DNA Chromatography analyses of RUSH alternative splicing [23], when they stated that the two isoforms differed by a 57 nt insertion (the RUSH 1α RT-PCR product was 225 bp, and the RUSH 1β 282 bp). This insertion introduced a stop codon that prematurely truncated the RUSH 1β isoform. This approach was subsequently used for measuring the relative levels of RUSH alternative splicing in a range of tissues.

This approach to detecting alternative splicing has been applied recently by others to thioredoxin-1 (Trx-1) in human cancers [24]. By combining competitive RT-PCR with DNA Chromatography analyses of the reaction products, these investigators were able to demonstrate that alternative splicing might exert some level of control over the amounts of Trx-1 found in certain cancer cells.

6.10
Differential Messenger RNA Display via DNA Chromatography

The development of methodologies for the analysis of mRNA populations in individual cells or tissues has revolutionized the investigation of gene regulatory networks in cell biology. Several methods are currently available for rapidly identifying both individual genes and sets of genes that are critical for developmental processes, or that mediate cellular responses. Such methods include differential display (DD) [4], comparative expressed sequence tag sequencing [25], representational difference analysis [26, 27], cDNA or oligonucleotide arrays [28, 29], and the serial analysis of gene expression [30]. DD has been shown to be a powerful approach for understanding the mechanisms of differentiation and development by detecting altered gene expression in closely related cell lines or tissues [4, 31–34]. The technique involves the detection of changes in the expression of mRNAs by their selective enrichment, without any prior knowledge of the sequence of the specific genes. In a typical DD experiment, total RNA is isolated from those cell types to be compared; first strand cDNAs are then synthesized by reverse transcription using an oligo-dT primer that has a specific dinucleotide at its 3′ end. This anchor primer and an arbitrarily chosen primer are then used in the PCR to amplify cDNAs to which both primers can hybridize [4].

The main drawbacks of DD are the lack of reproducibility, the inability to read and compare complex gels, and difficulties in recovering the desired bands from gels. Optimized primers and annealing temperatures can reduce the number of false positive results [35], while DNA Chromatography offers the potential of overcoming the problems related to comparing and recovering the bands from complex gels.

The experimental data shown in Figure 6.16, obtained from the human embryonal carcinoma cell line NTERA 2 (a model for human neuronal differentiation) [35], was used as a test bed system to establish whether DNA Chromatography could indeed replace the current gel-based steps used routinely in DD analysis. The use of this chromatographic approach permits multiple rounds of amplification, which may be especially helpful when the amount of isolated RNA is limited.

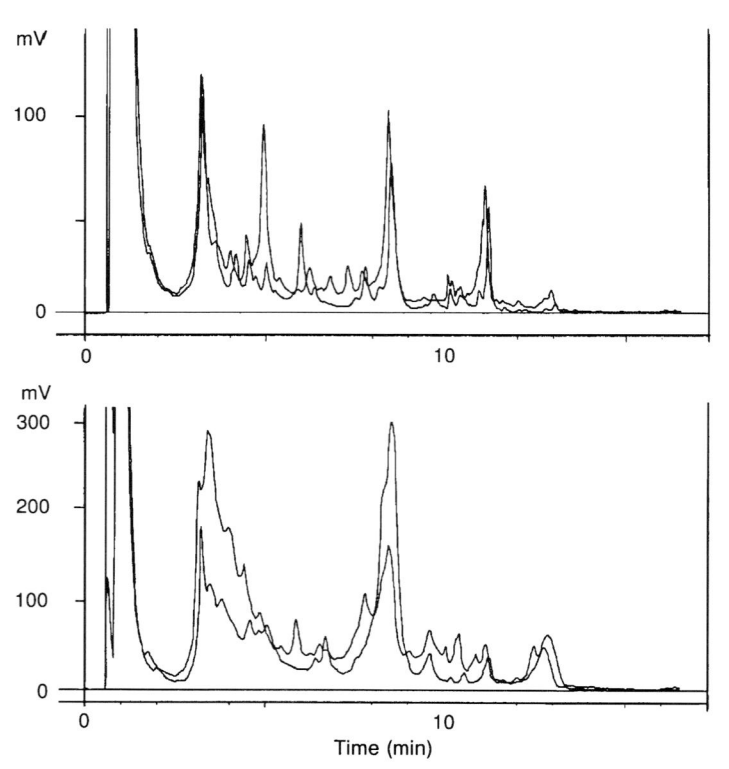

Figure 6.16 Two examples of differential chromatograms produced following RT-PCR of mRNA isolated from NTERA 2 cells, following the administration of retinoic acid. The differences in the traces arise through differential patterns of gene expression; the differences in the peaks can be related to specific differences in cDNA expression patterns by isolation of the peaks, cloning into a plasmid vector followed by nucleotide sequencing.

6.10 Differential Messenger RNA Display via DNA Chromatography

In addition, preparative chromatography simplifies the recovery and cloning of differentially expressed mRNAs as cDNAs. A schematic representation of chromatographic DD is shown in Figure 6.17.

NTERA 2 cells provided an excellent opportunity for validating the chromatographic approach to DD. Those genes which were shown to be differentially expressed in a preliminary screen were all either known to be expressed (or repressed) during retinoic acid stimulation, or carried the hallmarks of those clones of genes that would likely be differentially expressed in response to retinoic acid administration.

In comparison with the high-resolution micro-array analysis, with which it is possible to measure simultaneous fluctuations in the expression of thousands of individual mRNA species, chromatographic DD – as presented here – is less comprehensive. Thus, individual peaks in a DD chromatogram typically reflect subpopulations of mRNAs. Subsequent rounds of PCR-mediated peak interrogation can approach the resolution (but not the comprehensive coverage) of micro-arraying. However, given these limitations, it is the simplicity and low-cost nature of the DD technique that has been the prime motivation for its development. By contrast, whereas micro-array analysis can only be applied to the detection of changes in the expression of known genes, it is often desirable to detect and compare all mRNA species expressed in a particular cell, both known and unknown.

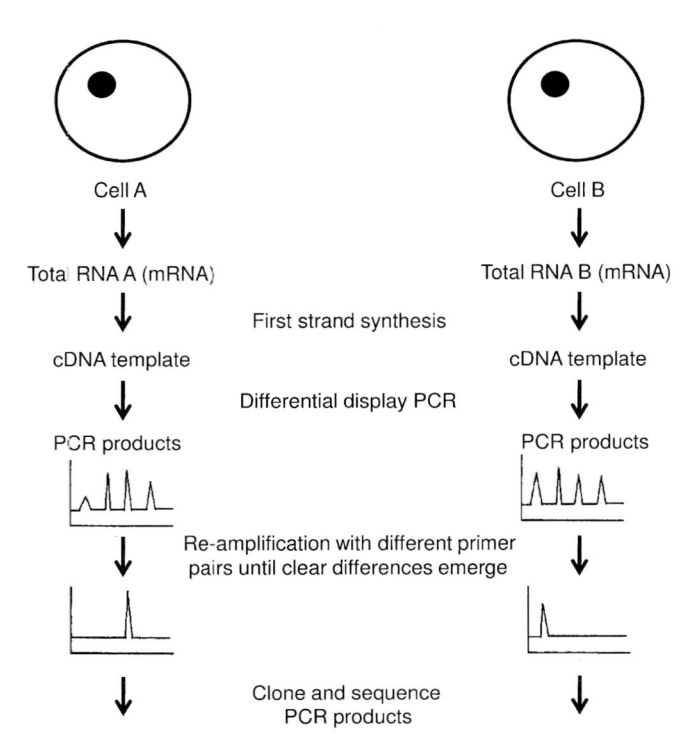

Figure 6.17 Schematic representation of the chromatographic differential message display analysis.

References

1 Conroy, M., Ashby, J.R. and Hornby, D.P. (1999) Patent Application, abandoned. University of Sheffield, Department of Molecular Biology and Biotechnology.
2 Gjerde, D.T., Hanna, C.P. and Hornby, D. (2002) *DNA Chromatography*, Wiley-VCH, Weinheim.
3 Waghmere, S.P., Pousinis, P., Hornby, D.P. and Dickman, M.J. (2009) Studying the mechanism of RNA separations using ion pair reverse phase liquid chromatography and its application in the analysis of ribosomal RNA and RNA:RNA interaction. *J. Chromatogr. A*, **1216** (9), 1377.
4 Dickman, M.J. (2005) Effects of sequence and structure in the separation of nucleic acids using ion pair reverse phase liquid chromatography. *J. Chromatogr. A*, **1076**, 83.
5 Dickman, M.J. (2007) Post column nucleic acid intercalation for the fluorescent detection of nucleic acids using ion pair reverse phase high-performance liquid chromatography. *Anal. Biochem.*, **360**, 282.
6 Dickman, M.J. and Hornby, D.P. (2006) Enrichment and analysis of RNA centered on ion pair reverse phase methodology. *RNA*, **12**, 691.
7 Latham, J.A. and Cech, T.R. (1989) Defining the inside and outside of a catalytic RNA molecule. *Science*, **245**, 276.
8 Fedor, M.J. (2000) Structure and function of the hairpin ribozyme. *J. Mol. Biol.*, **297**, 269.
9 Hampel, K.J., Walter, N.G. and Burke, J.M. (1998) The solvent-protected core of the hairpin ribozyme-substrate complex. *Biochemistry*, **37**, 14672.
10 Morasand, D. and Poterszman, A. (1996) Getting into the major groove protein-RNA interactions. *Curr. Biol.*, **6**, 530.
11 Dickman, M.J., Conroy, M., Grasby, J. and Hornby, D.P. (2002) RNA footprinting analysis using ion pair reverse phase liquid chromatography. *RNA*, **8**, 247.
12 Hampel, K.J. and Cowan, J.A. (1997) A unique mechanism for RNA catalysis: the role of metal cofactors in hairpin ribozyme cleavage. *Chem. Biol.*, **4**, 513.
13 Rupert, P.B. and Ferré-D'Amaré, A.R. (2001) Crystal structure of a hairpin ribozyme-inhibitor complex with implications for catalysis. *Nature*, **410**, 780.
14 Sambrook, J., Fritsch, E.F. and Maniatis, T. (1989) *Molecular Cloning: A Laboratory Manual*, 2nd edn, Cold Spring Harbor Laboratory Press, New York.
15 Oefner, P.J., Huber, C.G., Puchhammer-Stockl, E., Umlauft, F., Grunewald, K., Bonn, G.K. and Kunz, C. (1994) High-performance liquid chromatography for routine analysis of hepatitis C virus cDNA/PCR products. *Biotechniques*, **16**, 898.
16 Hayward-Lester, A., Oefner, P.J., Sabatini, S. and Doris, P.A. (1995) Accurate and absolute quantitative measurement of gene expression by single-tube RT-PCR and HPLC. *Genome Res.*, **5**, 494.
17 Hayward-Lester, A., Oefner, P.J. and Doris, P.A. (1996) Rapid quantification of gene expression by competitive RT-PCR and ion-pair reversed-phase HPLC. *Biotechniques*, **20**, 250.
18 Bodine, P.V., Green, J., Harris, H.A., Bhat, R.A., Stein, G.S., Lian, J.B. and Komm, B.S. (1997) Functional properties of a conditionally phenotypic, estrogen-responsive, human osteoblast cell line. *J. Cell Biochem.*, **65**, 368.
19 Doris, P.A., Oefner, P.J., Chilton, B.S. and Hayward-Lester, A. (1998) Quantitative analysis of gene expression by ion-pair high-performance liquid chromatography. *J. Chromatogr. A*, **806**, 47.
20 Hayward, A.L., Oefner, P.J., Sabatini, S., Kainer, D.B., Hinojos, C.A. and Doris, P.A. (1998) Modeling and analysis of competitive RT-PCR. *Nucleic Acids Res.*, **26**, 2511.
21 Doris, P.A., Hayward-Lester, A. and Hays, J.K., Sr (1997) Q-RT-PCR: data analysis software for measurement of gene expression by competitive RT-PCR. *Comput. Appl. Biosci.*, **13**, 587.

22 Hayward-Lester, A., Hewetson, A., Beale, E.G., Oefner, P.J., Doris, P.A. and Chilton, B.S. (1996) Cloning, characterization, and steroid-dependent post transcriptional processing of RUSH-1 alpha and beta, two uteroglobin promoter-binding proteins. *Mol. Endocrinol.*, **10**, 1335.

23 Robinson, C.A., Hayward-Lester, A., Hewetson, A., Oefner, P.J., Doris, P.A. and Chilton, B.S. (1997) Quantification of alternatively spliced RUSH mRNA isoforms by QRT-PCR and IP-RP-HPLC analysis: a new approach to measuring regulated splicing efficiency. *Gene*, **198**, 1.

24 Berggren, M.M. and Powis, G. (2001) Alternative splicing is associated with decreased expression of the redox proto-oncogene thioredoxin-1 in human cancers. *Arch. Biochem. Biophys.*, **389**, 144.

25 Adams, M.D., Kelley, J.M., Gocayne, J.D., Dubnick, M., Polymeropoulos, M.H., Xiao, H., Merril, C.R., Wu, A., Olde, B. and Moreno, R.F. (1991) Complementary-DNA sequencing–expressed sequence tags and human genome project. *Science*, **252**, 1651.

26 Lisitsyn, N.A., Lisitsyn, N.M. and Wigler, M. (1993) Cloning the differences between 2 complex genomes. *Science*, **259**, 946.

27 Hubank, M. and Schatz, D.G. (1994) Identifying differences in messenger-RNA expression by representational difference analysis of cDNA. *Nucleic Acids Res.*, **22**, 5640.

28 Ramsay, G. (1998) DNA chips: state-of-the art. *Nat. Biotechnol.*, **16**, 40.

29 Lipshutz, R.J., Fodor, S.P., Gingeras, T.R. and Lockhart, D.J. (1999) High-density synthetic oligonucleotide arrays. *Nat. Genet.*, **21**, 20.

30 Bauer, D., Müller, H., Reich, J., Riedel, H., Ahrenkiel, V., Warthoe, P. and Strauss, M. (1993) Identification of differentially expressed messenger-RNA species by an improved display technique (DDRT-PCR). *Nucleic Acids Res.*, **21**, 4272.

31 Miele, G., MacRae, L., McBride, D., Manson, J. and Clinton, M. (1998) Elimination of false positives generated through PCR re-amplification of differential display cDNA. *Biotechniques*, **25**, 138.

32 Martin, K.J. and Pardee, A.B. (1999) Principles of differential display. *Methods Enzymol.*, **303**, 234.

33 Andrews, P.W. (1984) Retinoic acid induces neuronal differentiation of a cloned human embryonal carcinoma cell-line in vitro. *Dev. Biol.*, **103**, 285.

34 Malhotra, K., Foltz, L., Mahoney, W.C. and Schueler, P.A. (1998) Interaction and effect of annealing temperature on primers used in differential display RT-PCR. *Nucleic Acids Res.*, **26**, 854.

35 Matin, M.M., Andrews, P.W. and Hornby, D.P. (2002) Multidimensional differential display via ion-pair reverse-phase denaturing high-performance liquid chromatography. *Anal. Biochem.*, **304**, 47.

7
RNA Structure–Function Probing

7.1
Definition of the Structure–Function Paradigm

The field of modern molecular biology grew out of two concepts that serve as the guiding principles that keep this vastly growing and often divergent field united. First, the "Central Dogma" of molecular biology states that DNA makes RNA, and RNA makes protein (Figure 7.1). This is a simplified model for the transfer of genetic information from a state of pure storage, DNA, to a state of pure activity, proteins. First proposed by Francis Crick in a seminar in 1958, the Central Dogma outlined the frontiers of biology and subsequently grew into the field of molecular biology.

Molecular biologists ask, namely, "How is DNA transcribed into RNA; how is RNA translated into protein; and how do proteins regulate life?" As the field advanced, many levels of regulation have been revealed that control DNA replication, RNA transcription, and protein translation. In fact, the Central Dogma is becoming increasingly complicated, and the mechanisms that regulate the transfer of information are still the subjects of active investigation.

The replication of DNA follows a "semi-conservative" process, in which the two strands of DNA that comprise a chromosome split to become templates for two newly synthesized DNA strands (Figure 7.2). The DNA strands are composed of a polymer of nucleotides – adenine, cytosine, guanine, and thymine. The hydro-

Figure 7.1 A schematic depiction of the flow of information representing the Central Dogma of molecular biology, first described by Francis Crick in 1958. The solid arrows state that the genetic information stored in DNA is transferred to RNA and the information in RNA is used to make protein. The circular arrow around DNA represents the process of replication, where the DNA itself is used as a template to replicate itself. The dashed arrow represents a special case in which the information stored in RNA is transferred to DNA. There are currently no known mechanisms for replication of information stored in a protein's sequence.

RNA Purification and Analysis: Sample Preparation, Extraction, Chromatography
Douglas T. Gjerde, Lee Hoang, and David Hornby
Copyright © 2009 WILEY-VCH Verlag GmbH & Co. KGaA, Weinheim
ISBN: 978-3-527-32116-2

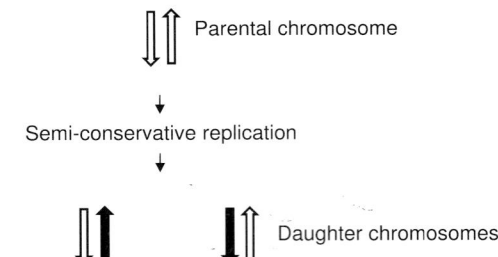

Figure 7.2 The process of DNA replication occurs via a semi-conservative process. The parental chromosome consists of two anti-parallel strands of DNA. The arrow pointing down represents the strand that travels in the 5′ → 3′ direction from top to bottom. The arrow pointing up is complementary to the first strand, and travels 5′ → 3′ from the bottom up. In semi-conservative replication, the two strands split (open arrows) and the cell's machinery, DNA polymerase, synthesizes a new strand of DNA (solid arrows) complementary to the parental strands. The resulting daughter cells will each contain a complete chromosome with one strand from the parent and one strand being newly synthesized.

gen-bonding potential and geometry of these nucleotides specifies a process of replication in which adenine base pairs to thymine, and cytosine base pairs to guanine, and are said to have complementarity. As a consequence, each daughter cell receives a chromosome consisting of an "old" DNA strand and a "new" DNA strand. Whilst many steps of the replication process are understood, it remains a mystery as to how a cell identifies the "old" strand, which eventually ceases to take part in replication. RNA replication is an equally enigmatic process.

It is understood that, during the process of transcription, the genetic code of DNA is copied into a working molecule of RNA called messenger RNA (mRNA). An organism's investment into the manufacture of a working copy of a gene may be a seemingly waste of resources; however, the transient mRNA molecule allows a cell to regulate the activity of its genetic code by regulating, first, the manufacture of mRNA, and second the regulation of mRNA degradation.

Higher eukaryotes introduce an interesting level of regulation in which some genes require processing after they have been transcribed into a mRNA. These RNAs are not fully mature mRNAs ready for translation into a protein, but rather are pre-messenger RNAs (pre-mRNAs), which contain noncoding sequences, termed introns, that are interspersed within the coding regions, termed exons. The introns may serve some other function in the cell as untranslated RNAs, they may have no known function, or they may code for parts of other proteins. The cell chooses how, and when, the introns are spliced out of the sequence by using a macromolecular complex made of RNA and protein, called the spliceosome (Figure 7.3). It is not yet fully understood how a cell chooses to include some exons, while excluding others, with the result being completely unique proteins. Alternative splicing serves to increase the diversity of the genome, not by increasing the number of genes, but by having multiple gene products from the same pre-mRNA. This is one explanation why the human genome is composed of a

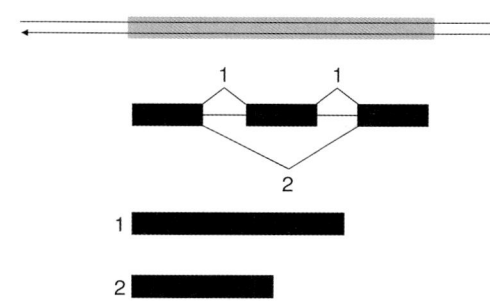

Figure 7.3 A gene (light gray box) is located in the chromosome. The cell transcribes the gene into a working copy of the gene called a pre-messenger RNA (pre-mRNA) that contains coding regions called exons (black boxes) and noncoding regions called introns (thin lines). In the process of splicing, the intervening sequences must be removed from the exons. When there is more than one way of splicing, the splicing machinery chooses which exons to include in the fully mature mRNA in a process called "alternative splicing". In scenario 1, all three exons are included; in scenario 2, only the first and last exons are included.

seemingly small number of genes relative to the genomes of less-complicated organisms.

In following the flow of genetic information from the DNA to the RNA, the next step is to pass from the RNA to the protein. At this level of gene regulation, the three-nucleotide codons specified by the mRNA sequence are translated into amino acids, the monomeric subunits that make up proteins. Only recently has gene regulation at the translational level been appreciated. The prevailing thought is that a tremendous amount of resources are invested into producing, regulating and splicing an mRNA, and that all mRNAs would be translated into proteins. The cell would regulate the translation of mRNA simply by degrading it when the appropriate level of protein had been reached.

The regulation of translation was previously thought to occur only at the level of initiation, where the translational machinery forms the initiation complex by binding to mRNA, to the initiator tRNA, and to the protein initiation factors. Up to this point, the cell has invested a great deal of energy to transcribe the RNA, splice out the introns and form the translation initiation complex. So, it came as quite a surprise to discover that another level of regulation existed where the mRNA, in the midst of being translated, could be signaled to stall the ribosome and stop translation. This level of regulation is carried out by short micro RNAs (miRNAs), which have some complementarity to the 3' end of the mRNA. These miRNAs bind to the mRNAs and form a structure which is recognized by protein components, including the RNA-induced silencing complex (RISC), the binding of which prevents the ribosome from translating any further. In fact, in some cases it will lead to degradation of the mRNA.

While the Central Dogma describes the flow of genetic information, the second concept guiding the field of molecular biology is the Structure–Function paradigm, which states that the three-dimensional (3-D) structure and shape of a biological molecule determines its cellular function. Many biologists consider a cell to be a bag of chemicals, and that it is only the regulation of the chemical reactions that separates a test tube from life. In order to regulate a cell's chemical reactions, however, a careful regulation of the formation and degradation of a cell's chemicals, as well as the control of each chemical reaction with respect to time, is required.

As outlined in this book, RNA has emerged as a catalytically active molecule, the formation, degradation and temporal control of which are regulated, along with proteins, to carry out the chemical reactions of a cell. The structure of the RNA is what determines its function; it is a molecule's structure that determines which chemical reaction takes place, and also ensures that the reaction occurs at the proper place, and at the proper time. The goal of RNA structural biology is to understand the RNA's structure, which will lead to an understanding of its function.

The rising interest in RNA structures correlates with the emerging cellular role of RNA, especially when biologists speculate on the details of the first biological molecule of relevance to appear in the prebiotic world. Although proteins are responsible for many processes in today's biological world, one problem persists in that proteins have no way of recording and replicating their chemical make-up, yet biology must have a means of reproducing. Such views form the basis for the RNA World Hypothesis, which is discussed later in the chapter.

RNA structure information is necessary for understanding the diverse roles that the molecule plays in all aspects of cellular function. Whilst the examination of RNA structures by themselves can solve many mysteries regarding RNA function, the RNA structure can, in the majority of cases, be rather complex, with the currently available tools providing little in correlating the shape of RNA with its function. In these instances, the RNA structure serves as a foundation upon which directed experiments can be based. This information is analogous to the determination of the human genome: just as the sequence information provides the basis to study the role of a gene, the structure of RNA provides a basis for the study of its mechanisms, dynamics, folding, and ligand interaction. Additionally, RNAs exist within the context of a cell, where they interact with many other molecules to form complexes and create interactions that specify many different functions. It is through a structural understanding of these interactions that the clues regarding the cellular function of these RNAs begin to emerge.

7.1.1
Francis Crick, and Predicting the Existence of tRNA

Following the discovery of DNA, Francis Crick made a series of seminal predictions. On examining the structure of DNA, he and James Watson deduced the fundamentals of DNA replication – that the anti-parallel, double-stranded DNA molecule could be replicated by way of complementarity to the nucleotides. Next, the structural principles that guide DNA replication were used to formulate more ideas about the translation step of the Central Dogma.

In a note to The RNA Tie Club in 1955, Crick predicted that an adapter molecule must be used to interpret genetic information contained in mRNA to make protein. As there was no structural evidence of complementarity between a nucleotide and an amino acid at the time, Crick articulated a prediction that must be true. His "adapter molecule" turned out to be transfer RNA (tRNA), which is used by a macromolecule – the ribosome – to make protein [1]. During the process of translating an mRNA into a protein, the ribosome polymerizes a polypeptide chain which is based on the codes specified by the mRNA, comprising sequential blocks of three nucleotides called codons.

Although today it is recognized that many tRNAs are used by the cell, each codon can be read only by a small subset of those tRNAs. Each tRNA codes for a single amino acid, with the tRNA being accurately aminoacylated to that a unique amino acid to create an aminoacylated tRNA (aa-tRNA). The ribosome then reads the mRNA by holding the mRNA in the correct reading frame and searching for the aa-tRNA which is complementary to the relevant codon. The aa-tRNAs are brought to the ribosome by protein elongation factors which prevent the aa-tRNA from completely entering the ribosome. Once a cognate codon/anticodon helix has been formed between the mRNA and the aa-tRNA, respectively, the ribosome triggers the elongation factor to hydrolyze a GTP molecule; this causes a conformational change that allows aa-tRNA to enter the ribosome. In this state, the amino acid bound to the tRNA is brought into the correct position relative to the growing peptide, so as to allow the peptide bond transfer that causes the peptide to be elongated by another amino acid. The function of the tRNA fits perfectly with Crick's adapter molecule hypothesis, which states that a cell requires a way to translate genetic information into protein information. The tRNA represents one of the earliest accounts of how the structure of an RNA determines its function, while the translational machinery will serve as an example of the RNA structure–function paradigm.

The crystal structure of tRNA was discovered, almost simultaneously, in two independent laboratories. The phenylalanine-specific tRNA from yeast (tRNA^{Phe}) was resolved at 2.5 Å and 3.0 Å [2, 3], and shown to be an L-shaped molecule containing certain key features worthy of examination. At one end of the molecule – the anticodon stem-loop-presented three nucleotides that could be "flipped out" of the loop, such that the Watson–Crick face of the nucleotides pointed into the solution. The other end of the molecule – the aminoacyl end – was shown to present free hydroxyl groups at positions 2′ and 3′ of the terminal ribose, at nucleotide 76 (Figure 7.4). A casual examination of this molecule would predict that the anticodon, in displaying three nucleotides, could be used to base pair with codons contained in the mRNA, in order to read the genetic code. The aminoacyl end of the tRNA provided the free substrates for attaching an amino acid. One end of the molecule was thought to decipher the genetic code, which was interpreted by the amino acid at the other end of the molecule. The specificity of the codon/anticodon helix ensured that the correct amino acid was delivered to the ribosome. The 76 Å distance separating the two ends of the tRNA molecule would predict that the ribosome contained two distinct sites – one for deciphering the genetic code, and one for carrying out the catalysis of peptide bond formation.

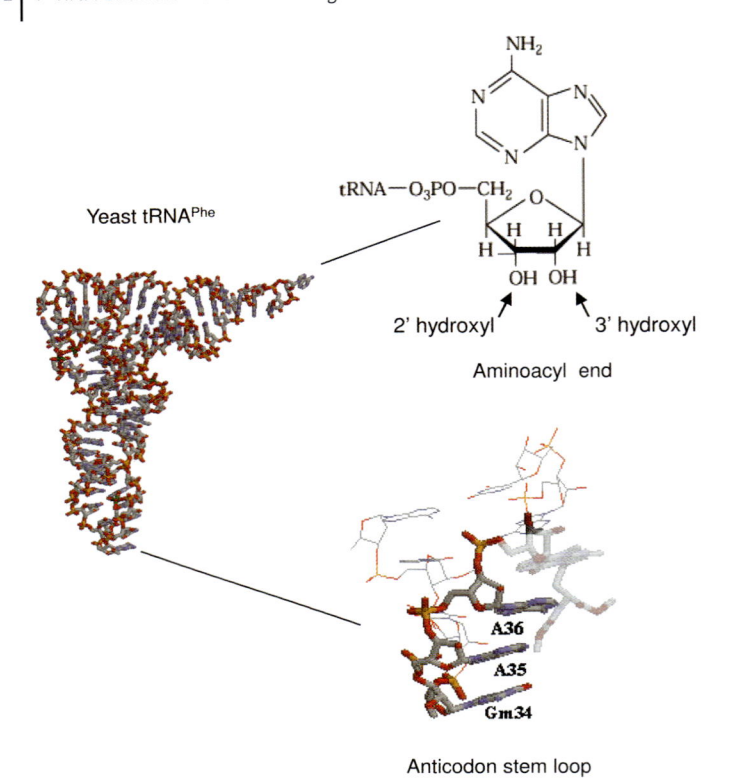

Figure 7.4 The structure of phenylalanine-specific tRNA from yeast. The L-shaped molecule features an aminoacyl end that binds to amino acids, and an anticodon that decodes mRNA. These two features are separated by more than 76 Å.

While a topical understanding of how the tRNA interacts with the ribosome could be deduced from an examination of the crystal structures of the tRNA molecule, the mechanism of aminoacylation—the process of attaching the appropriate amino acid onto the tRNA—was quite a surprise. Just as the anticodon provides the specificity used by the ribosome to decipher the genetic code, the aminoacyl tRNA synthetase (ARS) was thought to read the tRNA in the same way. Thus, a mistake in mis-aminoacylation of the tRNA would result in a mis-incorporation of amino acids into the protein, with subsequent detrimental effects to the cell.

The cocrystal structures of the ARS in complex with tRNA revealed that some ARSs do indeed read the anticodon, while others do not bind anywhere near the anticodon (Figure 7.5). It transpired that those features which were subtle to visual examination (because the structures of all tRNAs seem virtually the same) were chemically and proximally significant enough for the ARS to recognize them as a unique tRNA. In the same way that it is impossible to distinguish individuals among a herd of zebras, it is difficult for scientists to recognize the unique features of tRNAs.

tRNA^Gln-GlnRS co-crystal structure   tRNA^Ser-SerRS co-crystal structure

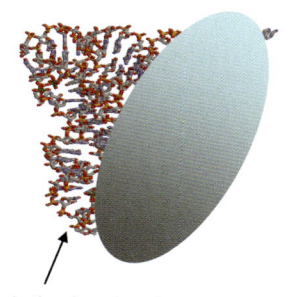

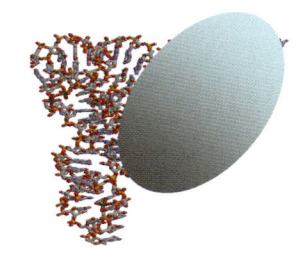

Anticodon stem loop

Figure 7.5 Cocrystal structures of tRNA bound to aminoacyl tRNA synthetases (gray ovals) reveal two modes of identifying the tRNA. Glycine aminoacyl tRNA synthetase recognizes features of the tRNA^Gln anticodon (left), while serine aminoacyl tRNA synthetase recognizes the backbone features of the tRNA^Ser that do not include the anticodon (right).

Structural information must be examined in the context of its biological significance, and also in the context of the whole cell. Thus, the conclusions and speculations drawn from a crystal structure must be examined by effective experimentation, including further structural studies, genetics, biochemistry, and cell biology.

Up to this point in time, the way in which a tRNA was considered to interact with the ribosome had been determined by scientific deduction. However, density gradient ultracentrifugation analyses of the ribosome – the cellular machinery responsible for translating mRNA into protein – showed the prokaryotic ribosome to be composed of two asymmetric subunits: (1) the small, 30S ribosomal subunit, which binds to mRNA; and (2) the larger, 50S subunit, which serves as the catalytic center for peptide bond formation. The mounting evidence that the anticodon stem–loop would bind to the small subunit, while the aminoacyl end would bind to the large subunit, was definitively confirmed by solving the crystal structure of the ribosome in complex with tRNAs (Figure 7.6) [4].

The 30S ribosomal subunit interacts with the anticodon stem–loop of the tRNA and the mRNA, thus confirming that this subunit could indeed read the genetic code. The information learned by the ribosome in the 30S ribosomal subunit is delivered to the 50S ribosomal subunit, where the peptide bond formation is catalyzed. It is likely that these long-range signal transmissions occur by way of conformational and/or chemical changes, although this proposal is currently under active investigation.

7.1.2
The Discovery of Ribozymes

Up until the late 1980s, RNA was thought to serve the cell as mostly a structural scaffold or as a temporary carrier of information. The 30-year period between 1950

7 RNA Structure–Function Probing

(a)

30S subunit

50S subunit

(b)

30S subunit

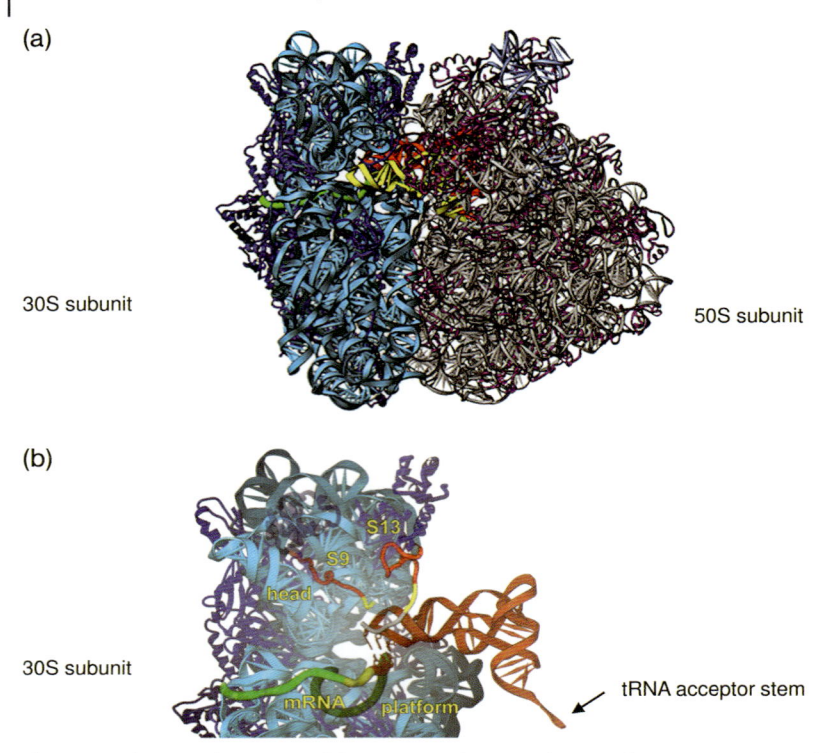

tRNA acceptor stem

Figure 7.6 The crystal structure of the ribosome in complex with three tRNAs and mRNA reveals how the different ligands interact with the ribosome. (a) The side view of the ribosome shows the small 30S ribosomal subunit and the large 50S ribosomal subunit completely encasing the tRNAs. (b) An enlarged view of the ribosome, with the 50S subunit and two tRNAs removed for clarity. The anticodon stem–loop of the tRNA forms contacts with the RNA and protein elements of the 30S subunit. The mRNA interacts with the tRNA anticodon and 30S subunit, exclusively. The acceptor stem interacts with the 50S subunit, and is the site of peptide bond formation.

and 1980 witnessed some 64 000 publications on the subject of RNA, the 10-year period between 1980 and 1990 saw 61 000 articles, and the 10-year period between 1990 and 2000 over 180 000 articles. The discovery of the catalytic activity of RNA by Tom Cech and Sidney Altman and colleagues had indeed started a flurry of interest in the study of RNA! [5, 6].

As scientists pondered the origins of life, research groups recognized that life was made up of polymers, namely DNA, RNA and protein. As a consequence, interest was directed towards how these monomers had been synthesized in a prebiotic world, how they would have been polymerized, and how they could have replicated so as to constitute life. DNA was of considerable interest here, because the mechanism for replication could explain how information would be passed on from one generation to the next; however, DNA itself could not carry out functions such as the creation of polymers from monomers. Strangely, proteins *were* able to catalyze these polymerization reactions, despite not being able to replicate themselves. This dilemma persisted for some time until the discovery that the *Tetrahymena* group I intron and the RNA portion of RNase P were in fact ribozymes, or RNA enzymes. The ability of RNA to perform cleavage reactions opened the door to the possibility that RNA could also carry out polymerization reactions. Here was a molecule, RNA, that carried genetic information, had a base-pair complementarity which suggested a mechanism of replication analogous to DNA replication, and had been shown to catalyze chemical reactions. This led to a hypothesis that RNA was in fact the precursor molecule in the prebiotic world, with the subsequently proposed "RNA World Hypothesis" going on to predict that life began with all organisms using RNA as the genetic and catalytic material [7]. The emergence of proteins, presumably, increased the efficiency of catalysis and added diversity and flexibility to the evolution of the organism.

The RNA World Hypothesis is difficult to prove, because no organism – modern or ancient – has been discovered to be composed exclusively of RNA, although much evidence exists which points towards RNA as the first prebiotic molecule. For example, some viruses use RNA, rather than DNA, as their carrier of genetic information, and employ the enzyme reverse transcriptase to copy the RNA genome into a complementary DNA (cDNA), which the host cell is tricked into replicating. In DNA replication, RNA primers are required for DNA replication, while the transcription of RNA requires no priming. RNA is composed of a ribose base, whereas DNA uses a deoxyribose base yet, metabolically speaking, ribose is the precursor to deoxyribose. Similarly, thymidylate synthase converts uracil, an RNA monomer, to thymine, a DNA monomer, which again suggests that the RNA is the precursor to DNA.

The study of RNA and RNA enzymes has a direct implication towards our understanding of an ancient world. Whilst a large body of evidence suggests that RNA is a precursor to DNA, the real clues as to how RNAs function in the prebiotic world are hidden in the mechanisms of RNA-driven catalysis. The discovery of these ribozymes and RNA aptamers – RNA molecules that can bind ligands – will provide an increasing amount of information – the challenge is to develop the tools and methods to fully characterize a once-overlooked molecule.

7.2
Structure Determination of RNA

When discussing RNA structure, it is useful to define the terms used to describe the molecule's structural information. RNA structures can be described on many levels, starting from the sequence and ending with a 3-D map of the interactions between molecules. The complexity of the structure has led research groups not only to simplify the representation of the structure but also to use smaller structures of discrete RNA domains to describe the more complicated structures.

7.2.1
Primary Structure

The primary structure is the sequence of nucleotide monomers, joined by a phosphodiester bond, that make up a specific RNA molecule. The monomer is composed of a base, ribose sugar, and phosphate group. Only the base is used in representing the monomer; hence, A = adenine; C = cytosine; G = guanine; and U = uracil.

An example of a primary sequence is:

```
            5            10           15           20
N–N–N–N–N–N–A–C–G–U–G–A–G–A–A–C–G–U–N–N–N–N–N–N
```

where N represents any nucleotide.

7.2.2
Secondary Structure

The secondary structure of an RNA molecule is defined as the base-paired arrangement of nucleotides that minimizes the free energy of the molecule. Typically, base pairs are first defined as Watson–Crick, after which additional non-W–C base pairs are allowed, according to observed pairing in known RNA structures. These may be noncanonical and may arise from alternative tautomeric forms.

It has been determined by both nuclear magnetic resonance (NMR) and X-ray crystallography, that ten consecutive base pairs will form a complete turn of a double helix. The above primary structure sequence would result in the following secondary structure, in which nucleotides 7 to 10 form a typical double-helical segment by base pairing with nucleotides 18–15, respectively.

```
            A G
           G   A
        10 U=A 15
           G=C
           C=G
        5         20
N–N–N–N–N–N–A=U–N–N–N–N–N–N
```

There are a number of assumptions to be made here, including that the partial helix adopts a regular A-form conformation. However, care should be taken to define a secondary structure strictly as the hydrogen bonding that results in the formation of helixes at the local level. It is easy to assign too much information to a secondary structure because of what is known about the tertiary structure. The conventional approach is to assume that a base-paired region is an A-helix, but it is possible that such regions can also adopt alternative conformations. The structure of the loop cannot be defined at the secondary level, nor can any 3-D information be concluded from a secondary structure.

Classically, the secondary structures are derived from an analysis of multiple sequence alignments. For example, the nucleotide sequence from a number of 16S ribosomal RNA-encoding genes was obtained from a large number of different bacterial species. These sequences could be aligned with confidence, as they are relatively constant throughout evolution. These regions were concluded to be functionally important, and therefore mutations in these regions would not be tolerated. Further analysis revealed that mutations immediately 5′ of these constant regions were consistent with mutations 3′ of the constant region, and that these mutations preserved a Watson–Crick complementarity. These mutations are said to be "covariant", and to form the basis for identifying base-paired helical regions of RNA. The fictional example below depicts such an analysis.

```
                5          10          15          20
N–N–N–N–N–N–A–C–G–U–G–A–G–A–A–C–G–U–N–N–N–N–N–N    E. coli
N–N–N–N–N–N–A–C–G–C–G–A–G–A–A–G–C–G–U–N–N–N–N–N–N    B. subtilis
N–N–N–N–N–N–G–C–G–C–G–A–G–A–A–G–C–G–C–N–N–N–N–N–N    T. thermophilus
N–N–N–N–N–N–U–G–G–C–G–A–G–A–A–G–C–C–A–N–N–N–N–N–N    S. dysenteriae
N–N–N–N–N–N–C–G–U–G–G–A–G–A–C–A–C–G–N–N–N–N–N–N    B. burgdorferi
```

The sequence from 11–14 is not tolerant of mutations across the different species shown. However, there is a high frequency of variation observed in the sequences surrounding 11–14. For position 7, the adenosine (A) in *E. coli* and *B. subtilis* is mutated to guanosine (G) in *T. thermophilus*. This A → G mutation is compensated by a U → C mutation at position 18, which preserves the Watson–Crick base pairing. As this trend is consistent throughout the regions surrounding the constant region, it can be concluded that the constant region is a loop and the surrounding regions form a helix; this is a stem–loop. The covariation rules have been expanded slightly to incorporate "wobble rules", under which G can base pair with U, and so on.

7.2.3
Tertiary Structure

The tertiary structure is defined as the complete 3-D structure of an RNA molecule and, like proteins, is defined by primary sequence information. The position of every atom in three dimensions must be known, as well as all of the bond angles, in order to define an unequivocal tertiary structure.

Traditionally, solving NMR or X-ray structures is the only way to reveal the tertiary structure of RNA. The 3-D structure of the above RNA would confirm predictions from secondary structure analysis, such as whether a specific region adopted a GNRA tetraloop configuration; this would also determine how the molecule is folded. Tertiary structures reveal interactions between distant structures;

for example, the structure may involve interactions between the above tetraloop with other loops, helixes or sequences elsewhere in the molecule, as is the case with polypeptides.

Some examples of tertiary structures are as follows:

- A *stem–loop* (also known as a hairpin or hairpin loop) is a contiguous stretch of nucleotides that possesses little or no hydrogen-bonding potential. Loops may interact with distant structures, but cannot interact with nucleotides in close proximity. The stem is the helical region of Watson–Crick base pairing formed at the immediate 5′ and 3′ regions of the loop; the stem is a helical region of at least one base pair, and can be of indefinite length. The helical regions must be continuous, and may comprise wobble base pairs.

- *Tetraloop bulges* are helical regions that are not continuous. They are found when canonical or wobble base pairing are not observed at one base pairing position, or when the 5′ strand is one nucleotide longer than the 3′ end and *vice versa*, so that one nucleotide is unable to base pair within the helix.

- *Internal loops* are a form of bulge in which more than one adjacent nucleotide do not base pair to form a helical segment.

- *Pseudoknots* comprise two stem–loops in which each loop region forms the helical stem of the other stem–loop.

The structures listed here, along with helical structures, comprise the building blocks for RNA structures of higher complexity. For simplification, these building blocks are represented at the secondary level by depicting the hydrogen bonding of the helical segments and leaving the bulges and loops unpaired. When these building blocks are found in 3-D structures, they may be independent of tertiary interactions, in which case their shape is predictable. If these building block structures form tertiary interactions, the shapes of the building blocks will be distorted. These distortions include, for example, the hydrogen bonding of loop and bulge regions, base stacking, minor groove interactions, and salt bridges.

7.2.4
Quaternary Structure

Quaternary structure is defined as interactions between more than one tertiary structure. An example of this would be the various interactions between a tRNA and a ribosome.

The most common methods for determining RNA structure are X-ray crystallography, NMR and cryo-electron microscopy (cryo-EM). The Protein Data Bank reports structures of over 45 000 proteins, but fewer than 2000 RNAs and RNA–protein complexes, which highlights the difficulties of RNA structure determination. The building blocks of RNA, the ribonucleotide triphosphates, are quite large compared to the amino acids, the building blocks of a protein. This large size and complexity of the RNA monomers limits the sizes of RNAs that can feasibly be resolved with NMR. Crystal structures are difficult to obtain because the electronegative RNA backbone reduces the ability of the RNA to form a crystal. Cryo-EM has been quite successful in determining the 3-D structures of RNA, although these structures tend to be of a lower resolution. The cryo-EM structure-determining process also lacks internal standards with which to measure the accuracy of a structure, unlike NMR and X-ray crystallography. While biology clearly requires the details of RNA structural information, traditional methods have been supplemented by some newer methods.

7.3
Footprinting, Model Building, and Functional Investigations

Techniques have been developed to obtain structural information without using traditional approaches. The information obtained may serve as a useful alternative for structure determination, notably due to the relative speed and ease of these techniques, compared to traditional approaches. These methods may also complement low-resolution structures by providing additional valuable information. The techniques generate proximity information by identifying the nucleotides around a known position, and include chemical probing, nucleotide analogue interference mapping (NAIM), hydroxyl radical probing, directed hydroxyl radical probing, and Förster resonance energy transfer (FRET).

RNA footprinting owes its roots to the investigations conducted to determine protein–DNA interactions. An exploration of DNA footprinting provides the foundation for RNA footprinting, since the regulation of expression and the introduction of structural modifications to the genome rely on proteins that bind to specific DNA sequences. Understanding how proteins recognize their binding sites in the midst of a plethora of similar sequences in the genome, represents an important step in identifying how these processes function at the molecular level.

The elegant method of "footprinting" involves the binding of a protein to a radioactively labeled DNA which contains the sequence that the protein recognizes. This complex is then digested, either enzymatically using DNase I or chemically using hydroxyl radicals. Those regions of the DNA molecule which are covered by the bound protein will be protected from digestion, while the remainder of the DNA backbone will undergo cleavage. The products of the reaction are then separated using electrophoresis, with the extent of digestion being quantified by further analysis of the gel. The blank region on the autoradiograph, which corresponds to the site of the protein binding to the DNA, is termed the "footprint".

DNA footprinting techniques have been adapted to RNA footprinting with only slight modifications, in a bid to answer further questions relating to RNA quaternary interactions. If the structure of an RNA is has been solved, this technique can be applied to map position(s) on the RNA where various ligands would bind.

Footprinting can be a very fast way of obtaining these binding maps. With traditional methods, the only way to obtain binding maps would be to solve the crystal structures of the RNA–protein complex. If the determination of single structures is deemed to be difficult, then the determination of cocrystal structures would be even more so. However, once the structure of the bacterial ribosome had been solved, models proposing how the ribosome would interact with ligands were quick to emerge [8, 9].

The bacterial ribosome is a protein–RNA complex composed of over 50 proteins and three RNA molecules totaling over 4500 nt, and with a total overall molecular weight of 2.5×10^6 Da. By using the chemical differences between RNA and protein, a chemical approach could be developed to map the interaction between the ribosomal RNAs and ribosomal proteins and protein factors. Once the structures of the ligands had been deduced – in this case IF3 (a protein translation initiation factor) and RRF (a protein involved in ribosome recycling) – then fairly accurate models could be determined.

7.3.1
Chemical Probing and Cleavage

A structured RNA utilizes Watson–Crick base pairing, electrostatic interactions, van der Waals interactions, hydrogen bonding, and other weak interactions to create folding of the molecule. An understanding of this folding, and also how proteins, other RNAs and ligands interact with RNA, can be acquired by carrying out chemical probing and footprinting. For this, the investigators use a set of RNA modification agents which probe the structure of the RNA by covalently binding to accessible nucleotides. The modified nucleotides are identified by primer extension analysis, after which the RNA is extracted from the probing reaction and a DNA primer annealed to the 3′ end of the RNA. An enzyme, reverse transcriptase, is then used to extend the primer in the presence of radiolabeled deoxyribonucleotide triphosphates (dNTPs) to produce a cDNA molecule that is then analyzed by polyacrylamide gel electrophoresis (PAGE). The reverse transcriptase cannot read through positions that have been modified, which are visualized by a faster-migrating band on the gel (Figure 7.7). A correct probing reaction must be carefully optimized so that each RNA molecule is modified at approximately one position:

CMCT modifies U at N-3
CMCT modifies G at N-1
DEPC modifies A at N-7
DMS modifies A at N-1
DMS modifies C at N-3
Kethoxal modifies G at N-1 and C-2

When using this technique to understand folding, it is necessary to compare those nucleotides that are susceptible to modification when the RNA is denatured to those that are susceptible when the RNA is folded. The folded RNA will result

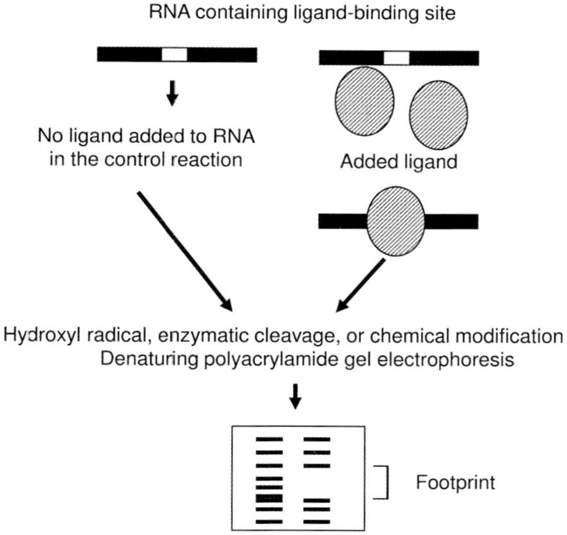

RNA containing ligand-binding site

Figure 7.7 Footprinting experiments can be used to map the binding positions of ligands to specific nucleotides of an RNA. A probing agent such as a nuclease, hydroxyl radical, or chemical modification agent is introduced to *in vitro* binding reactions of an RNA in the presence or absence of ligand. Following RNA modification, the RNA is extracted and the positions of modification are measured by primer extension and polyacrylamide gel electrophoresis. Comparison of the control (no ligand present) lane with the experimental lane (ligand present) reveals an area protected from modification called the ligand's "footprint".

in the loss of modification at positions where the folding event protects positions from modification. This protection occurs when the accessibility of that nucleotide is impeded for a number of reasons: (1) that the position participates in a hydrogen-bonding event; (2) that the position becomes buried in the core of the molecule; and (3) that the position is protected by a secondary event such as binding to an ion, metal or water molecule present in the buffer.

This analysis can offer the first clues as to how an RNA is folded, but without solving a structure. The positions that are modified in the folded structure must be exposed to solvent, and may be located at the periphery of the molecule, whereas the protected positions may be located in the core of the molecule. These experiments provide information as to how the RNA is constrained, and may be especially useful when combined with other model-building techniques.

Chemical probing can be used to determine the areas that are involved in ligand binding. If a solvent-accessible nucleotide participates in hydrogen bonding to a ligand, then it would not be susceptible to modification by the chemicals described above. In this way, the ligand puts a "footprint" on the RNA. This logic can then be used to determine which nucleotides on the RNA are protected from modification upon ligand binding, and to infer that the ligand protects those positions from modification through a direct binding event (Figure 7.8). It should be noted that

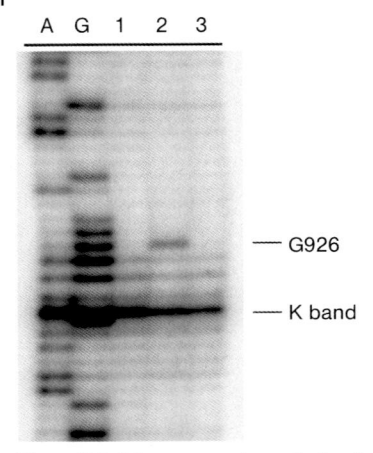

Figure 7.8 Primer extension gel of a ribosome footprinting experiment. Lanes A and G are sequencing lanes used to orient to the area of interest. Lane 1 is RNA-only, not treated with the modification agent, and is used to monitor positions that have a propensity for reverse transcriptase to stop (K bands). Lane 2 is the no-ligand control lane modified with kethoxal, and is used to detect positions that are modifiable (position 926). Lane 3 is the binding reaction of ribosome, mRNA, and tRNA. The loss of modification at position 926 in lane 3 suggests that tRNA footprints to that position.

these protections are only inferences, and that other events may lead to the protection of these positions. For example, binding of the ligand can be an indirect protection. Instead of the ligand binding to the position to cause protection, a conformational change in the folding of the RNA may have occurred upon binding of the ligand. This conformational change could result in protection of that position. Additionally, binding the ligand could induce binding of a small molecule, an ion, a metal or a water, for example, present in the reaction buffer that would protect that position from modification.

An extension of the chemical probing techniques is the induction of a cleavage event and the analysis of the protection of those events. The advantage of cleavage probing compared to chemical modification probing is that cleavage probing addresses backbone interactions whereas chemical modification probing addresses the hydrogen bonding activity of the Watson-Crick face of the nucleotide. Cleavage can be induced by using non-specific ribonucleases (RNases) such as RNase T1 or RNase A.

Chemical cleavage is another method used for RNA cleavage probing. It is possible to generate hydroxyl radicals by supplementing a typical biological buffer with hydrogen peroxide and iron. The hydroxyl radicals will abstract the proton at the 2′ ribose position, leaving an electronegative oxygen. The electronegative 2′ oxygen resolves its activated state by attacking the ester-linked phosphorus at the ribose 3′ position, with the 3′ end of the RNA acting as a very good leaving group. As before, primer extension analysis can be used to analyze cleavage of the RNA. Additionally, the RNA can be radioactively end-labeled and examined directly with

PAGE. This type of probing can be used to study RNA folding in order to generate structural information, and ligand binding or conformation changes can also be measured in this way. Often, these data are combined with chemical modification probing data to generate further details about folding or binding events. Notably, more data results in a more accurate model.

An extension of the cleavage probing analysis of RNA is that of directed hydroxyl radical probing. The generation of hydroxyl radicals can both direct and restrict cleavage events to specific locations on the RNA, but is limited to the analysis of RNAs that bind to proteins. The protein may be a recombinant protein in which each construct codes for a single cysteine mutant engineered at strategic points along its length. The cysteine is modified by a molecule, 1-(*p*-bromoacetamidobenzyl)-EDTA (BABE), to which an Fe(II) is tethered. The iron generates hydroxyl radicals of limited range, so that the RNA neighborhood at the iron is susceptible to cleavage.

A further advance of this hydroxyl radical cleavage technique allows for kinetic analysis. As the structure of the RNA determines its function, changing this structure may represent a means of controlling the functional events. In other words, RNA molecules are dynamic and will undergo movements as they transition from one state to another, although these transient events occur very quickly and are difficult to capture. When possible, crystal structures of a molecule in different catalytic states are determined, but these structures are quite rare. As certain conditions allow for controlled hydroxyl radical probing, it is possible to elucidate the transition states of an RNA chemical reaction; here, the use of synchrotron X-rays to generate hydroxyl radicals in combination with quench flow mixers allows for time-resolved experiments on the millisecond time scale [10].

Another approach for defining the area of cleavage is to use engineered nucleotide analogues at specific positions on a synthetic RNA. By strategically placing 5'-*O*-(1-thio)-nucleoside analogue triphosphates into the RNA, the sample can be treated with iodine such that cleavage will occur at the analogue nucleotides [11]. This technique can also be used to determine whether cleavage at that position is detrimental to the function of the RNA.

7.3.2
Modification Interference

If a ligand is known to bind to RNA, this technique seeks to interfere with that binding reaction in order to infer the importance of nucleotides in the RNA. The ligand is first immobilized onto agarose support beads, which are packed into a column. The active and properly folded RNA is added to the column before and after treatment (see Section 7.3.1) to compare a positive binding event against disruption of the binding. Those RNAs that are unable to bind to the ligand are washed through the column and collected. Primer extension analysis can be used to identify which nucleotides, when modified, will interfere with binding to the ligand [12].

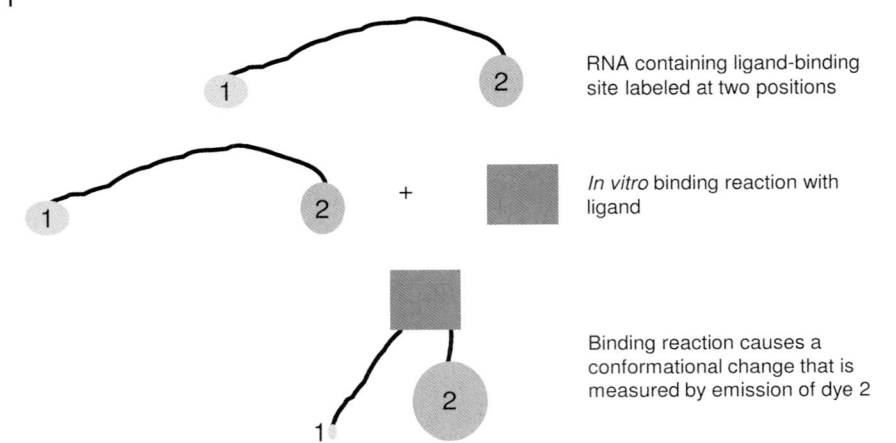

Figure 7.9 A FRET experiment is used to measure the conformational changes that occur on ligand binding. An RNA containing a ligand-binding site is labeled with two different dyes, 1 and 2. When the dyes are apart, the excitation and emission from the dyes are monitored as a baseline. On binding to the ligand, a conformational change occurs to bring the dyes in close proximity. The dyes are chosen such that the emission wavelength of dye 1 is the excitation wavelength of dye 2. Monitoring the increase of emission from dye 2 is directly proportional to the small distance separating the two dyes.

7.3.3
FRET

Today, it is increasingly apparent that RNAs are not only rigid structures but are also dynamic. The FRET technique involves the attachment of two fluorescent dyes to different positions on the RNA. The two dyes are selected such that an optimal distance between them will allow the emission wavelength of one dye to activate the excitation of the other dye. As the RNA moves, the emission of the second dye is measured and used to calculate the distance by which the two dyes have moved relative to each other (Figure 7.9) [13].

References

1 Hoagland, M.B., Zamecnik, P.C. and Stephenson, M.L. (1957) Intermediate reactions in protein biosynthesis. *Biochim. Biophys. Acta*, **24**, 215.
2 Kim, S.H., Suddath, F.L., Quigley, G.J., McPherson, A., Sussman, J.L., Wang, A.H., Seeman, N.C. and Rich, A. (1974) Three-dimensional tertiary structure of yeast phenylalanine transfer RNA. *Science*, **185**, 435.
3 Robertus, J.D., Ladner, J.E., Finch, J.T., Rhodes, D., Brown, R.S., Clark, B.F. and Klug, A. (1974) Structure of yeast phenylalanine tRNA at 3 A resolution. *Nature*, **250**, 546.
4 Cate, J.H., Yusupov, M.M., Yusupova, G.Z., Earnest, T.N. and Noller, H.F. (1999) X-ray crystal structures of 70S ribosome functional complexes. *Science*, **285**, 2095.

5 Kruger, K., Grabowski, P.J., Zaug, A.J., Sands, J., Gottschling, D.E. and Cech, T.R. (1982) Self-splicing RNA: autoexcision and autocyclization of the ribosomal RNA intervening sequence of Tetrahymena. *Cell*, **31**, 147.

6 Guerrier-Takada, C., Gardiner, K., Marsh, T., Pace, N. and Altman, S. (1983) The RNA moiety of ribonuclease P is the catalytic subunit of the enzyme. *Cell*, **35**, 849.

7 Gesteland, R.F., Cech, T.R. and Atkins, J.F. (2008) *The RNA World*, 3rd edn, Cold Spring Harbor Laboratory Press, Cold Spring Harbor, New York.

8 Dallas, A. and Noller, H.F. (2001) Interaction of translation initiation factor 3 with the 30S ribosomal subunit. *Mol. Cell*, **8**, 855.

9 Lancaster, L., Kiel, M.C., Kaji, A. and Noller, H.F. (2002) Orientation of ribosome recycling factor in the ribosome from directed hydroxyl radical probing. *Cell*, **111**, 129.

10 Nguyenle, T., Laurberg, M., Brenowitz, M. and Noller, H.F. (2006) Following the dynamics of changes in solvent accessibility of 16S and 23S rRNA during ribosomal subunit association using synchrotron-generated hydroxyl radicals. *J. Mol. Biol.*, **359**, 1235.

11 Ryder, S.P. and Strobel, S.A. (1999) Nucleotide analog interference mapping. *Methods*, **18**, 38.

12 von Ahsen, U., Green, R., Schroeder, R. and Noller, H.F. (1997) Identification of 2′-hydroxyl groups required for interaction of a tRNA anticodon stem-loop region with the ribosome. *RNA*, **3**, 49.

13 Blanchard, S.C., Kim, H.D., Gonzalez, R.L., Jr, Puglisi, J.D. and Chu, S. (2004) tRNA dynamics on the ribosome during translation. *Proc. Natl Acad. Sci. USA*, **101**, 12893.

Appendix 1
Chromatographic Separation Equations and Principles for RNA Separation

A1.1
Basic Chromatographic Considerations

The process of chromatography produces peaks of sample materials that are separated into discrete bands. An optimized process of chromatography produces as many discrete sample bands as possible, that are separated in as short a time as possible. The implementation of a successful chromatographic process can easily be performed by using standard software and "cookbook" operational procedures found in the literature, and also in the application notes provided by instrument and column manufacturers. However, the optimization of a successful separation requires that chromatographers have at least some working knowledge of the theoretical and practical aspects of the chromatography column and HPLC instrumentation.

While this appendix provides a theoretical basis for chromatography, Appendices 2 and 3 further practical, working information for chromatographic RNA separations. Equations have been included in Appendix 1 because the use of mathematics can help the chromatographer develop new methods, or optimize existing methods, by understanding which parameters are important to a separation and then optimizing those parameters. Although many of these equations may be too advanced for the beginner, an attempt should be made to read this appendix anyway, as many of the concepts can be understood and used. And, as the user becomes more skilled and advanced, the appendix can be revisited and other chromatographic texts read, understood, and used.

A1.1.1
Retention

The major chromatography terms that describe performance are summarized in Figure A1.1 and Table A1.1. The derivation and use of these terms are described in the next few sections. Although the use of these equations and terms is not necessary for the beginner, their value is more apparent for those users who develop or optimize methods.

RNA Purification and Analysis: Sample Preparation, Extraction, Chromatography
Douglas T. Gjerde, Lee Hoang, and David Hornby
Copyright © 2009 WILEY-VCH Verlag GmbH & Co. KGaA, Weinheim
ISBN: 978-3-527-32116-2

Appendix 1 Chromatographic Separation Equations and Principles for RNA Separation

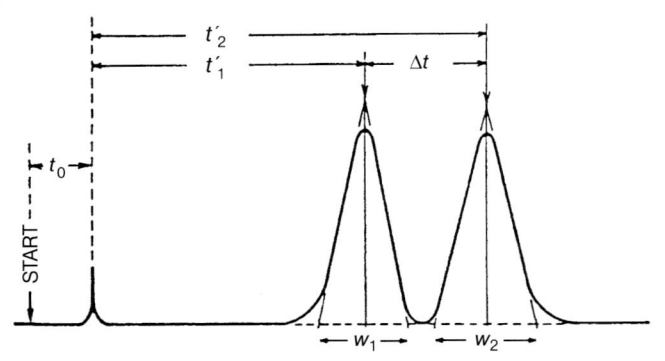

Figure A1.1 Two chromatographic peaks and their relationship to the chromatographic terms. The retention time, t, is the sum of the void time, t_0, and the adjusted retention time, t'. See Table A1.1 for a listing of terms and definitions.

Table A1.1 Chromatographic terms.

Term	Symbol	Definition
Retention time	t or t_R	Time to elute a peak to its maximum height
Dead time	t_0 or t_M	Time to elute a nonsorbed marker
Adjusted retention time	t'	$t' = t - t_0$
Retention factor (or capacity factor)	k or k'	$k = \dfrac{t - t_0}{t_0}$
Peak width	w	Peak width at its base, usually in time units
Peak width at 1/2 peak height	$w_{1/2}$	Peak width at half-height, usually in time units
Peak resolution	R_s	$R_s = \dfrac{2\Delta t}{\text{peak 1 W} + \text{peak 2 W}}$
Separation factor	α	$\alpha = \dfrac{k_2}{k_1} = \dfrac{t'_2}{t'_1}$
Theoretical plate number	N	$N = 16\dfrac{(t)^2}{W} = 5.54\dfrac{(t)^2}{W_{1/2}}$
Height equivalent of a theoretical plate	H	$H = \dfrac{L}{N}$ (L = column length)

During their passage through the column, sample molecules spend part of the time in the stationary phase, and part in the mobile phase. All molecules spend the same total amount of time in the mobile phase; this time is called the column dead time or holdup time, t_0. This time is measured by injecting a completely unretained compound and measuring the time required for the peak (measured at peak height) to reach the detector. The retention time, t, is the time from when the sample is introduced to the column to when the detector senses the maximum of the retained peak. This value is greater than the column holdup time by the amount of time that the compound has spent in the in stationary phase, and is called the adjusted retention time (t'). These values lead to the fundamental relationship describing retention time:

$$t = t' + t_0 \tag{A1.1}$$

or as it is usually expressed:

$$t' = t - t_0 \tag{A1.2}$$

Some refer to this quantity as the dead time, t_0, while others call it the time of mobile phase to pass, t_M. Although the retention time (the time for a sample component to be eluted from the column to its peak maximum) has been given the symbol t_R, we prefer to simply write it as t, as this makes it possible to denote the retention times of several peaks as t_1, t_2, t_3, ...

Retention is usually measured in units of time, but may also be measured in volume. The volume of eluent required to elute a substance from a column to its peak maximum is called the retention volume, V. With the conversion equation, volume can be substituted for time for any equation:

$$V = t \times F \tag{A1.3}$$
$$(\text{ml}) = (\text{min}) \times (\text{ml min}^{-1})$$

where F is the volumetric flow rate and

$$V = V' + V_0 \tag{A1.4}$$

or

$$V' = V - V_0 \tag{A1.5}$$

A1.1.2
Retention Factors

Perhaps the single most important term in liquid chromatography is the retention factor (or capacity factor), k.

$$k = \frac{t - t_0}{t_0} \tag{A1.6}$$

or

$$k = \frac{V - V_0}{V_0} \tag{A1.7}$$

Currently, it is recommended to use the term *retention factor* for what for many years was called the *capacity factor*. Both k and k' have been used as the symbol for this term.

Conditions must be adjusted so that there is a sufficient difference in the k-values of the various compounds in the mixture so as to provide a good separation from each other. It is also necessary to select conditions so that the range of k-values for each individual compound is such that a separation may be completed within a reasonable time. A range of k-values from 2 to 10 has often been suggested as the most desirable.

A1.1.3
Peak Width

Figure A1.2 shows the various measurements of peak width. The chromatographic peak is assumed to be symmetrical and Gaussian-shaped. However, as this is not true for real peaks, due to peak tailing, it is easier to measure the peak width at 10%

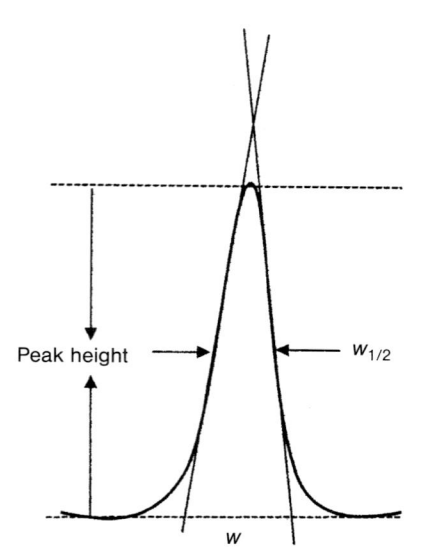

Figure A1.2 The peak width can be measured in different ways. The most preferred is measurement of the peak width at half-height. An estimation of peak width can be made: $w = 1.7 \times w_{1/2}$.

height or at ½ peak height. Retention and peak width are used to calculate various performance measurements; in this way, the separating power of a system can be calculated and compared. However, in order to do this, the chromatographer must rely on two fundamental theories of chromatography: (i) the plate theory; and (ii) the rate theory. The theories presented in this book are quite basic and have been simplified for the purpose of explaining the fundamental procedures. More detailed and expanded theoretical approaches can be found elsewhere [1–9].

A1.2
Plate Theory of Chromatography

In effect, it can be said that a chromatographic column provides for a large number of simultaneous equilibrations of sample material between the mobile and stationary phases. The plate theory of chromatography first attempts to define the chromatographic process through equations that describe a single equilibration of a single sample between the stationary and mobile phases. Later, the equations are expanded to assume multiple equilibrations in a column separations. Each mathematical equilibrium is said to be a theoretical plate.

Much of the early chromatographic studies were conducted using column packing materials where an insoluble stationary liquid phase was coated onto the solid packing material. In this case, it can be said that the sample material partitions between two liquid phases – one stationary, and the other mobile. For this reason, we shall call the sample compound a "solute". The retention volume of solute A eluting from a column can be calculated. A concentration distribution coefficient, D_C, is defined as:

$$D_C = \frac{[A]_S}{[A]_M} = \frac{(\text{mmoles A})_S/V_S}{(\text{mmoles A})_M/V_M} = D_m \frac{V_S}{V_M} \tag{A1.8}$$

Here, the subscripts S and M refer to the stationary and mobile phases, respectively; and $[A]_S$ and $[A]_M$ are the concentrations of the solute A in each phase. V is the volume of a liquid phase and V_S and V_M are the volumes of liquid in each phase. D_m is the mass distribution ratio (the ratio of the amount of solute in the stationary phase to the amount of solute in the mobile phase).

Assuming that solute retention is defined by the relative amount of solutes in the two phases, it can also be seen that the following equations describe the elution volume of a solute. Taking Equation A1.4 from above, V_M may be substituted for V_0, the dead volume for a non-retained fragment. (This makes the assumption that all of the dead volume is within the column, which is not exactly true because there is some dead volume within the tubing and connections; however, the values are near enough for our purposes here.) The increase in elution volume is directly proportional to the ratio of the solute in the stationary phase over the mobile phase, multiplied by the volume of the mobile phase. Thus, D_m multiplied by V_M may be substituted for V'.

$$V = V_M D_m + V_M \tag{A1.9}$$

Another way of expressing this through rearrangement is:

$$D_m = \frac{V - V_M}{V_M} \tag{A1.10}$$

Equation A1.10 is quite similar to Equation A1.7 above, since V_0 and V_M are for our purposes said to be the same. Combining Equation A1.8 with Equation A1.10 gives the following:

$$V = D_c V_s + V_M \tag{A1.11}$$

D_c can be measured in a single equilibrium experiment, where the column packing material is equilibrated with a known volume of solvent and a known concentration of solute. Measurements are made of the initial solute concentration and equilibrated solute in solution. Assuming that all of the solute that has gone from solution is now on the column packing material, the D_c can be calculated. It should be noted that D_m can also be calculated either with single equilibrium equations or by using column parameters. Once the retention volume has been determined, the retention time is calculated using the flow rate.

Of course, the separation of two or more compounds requires conditions where the distribution ratios are different. But, how different must they be to achieve complete separation? That depends on two parameters: (i) the difference of the distribution ratios; and (ii) the performance of the system. One method to achieve separation is to choose conditions where the distribution ratios are very different; here, materials with a low distribution ratio will elute early in the chromatogram, while those with larger distribution ratios will elute later. Although this may be quite effective, larger distribution ratios will increase the retention times so much that if they may become too large and it may take hours for a peak to elute from the column. In cases where the distribution ratios are quite different, a gradient must be used to persuade the compounds to elute in a reasonable (short) time. The use of gradients will be discussed later. The second method of achieving separation is to use a column system that has a high number of theoretical plates.

Recall, that the plate theory assumed only a single equilibration or one theoretical plate to predict the retention time of a compound. As the number of theoretical plates is increased in a column, the peak becomes narrower, and narrower peaks allow the separation of more compounds within one run. Fortunately, a flowing column system has many column interactions of the same type, which means that the number of theoretical plates of a column can be quite large. Extremely high, theoretical plate numbers have been reported in the literature, of up to 1 000 000, although in most cases a chromatographic system will have between 2000 and 15 000 theoretical plates with which to work.

So, there are really two ways to accomplish a separation. The first way is to choose conditions so that there is a large difference in D_c (which means there is also a large difference in the retention factor k). The second way is to use a system that has a large number of theoretical plates, and can therefore tolerate very small differences in retention factors. Most chromatographers use N (or sometimes n) to denote the number of theoretical plates, which is used as a measure of the separation power of a particular chromatographic column and system. A larger N indicates a greater resolving or separation power of the column in that particular system. The height equivalent of a theoretical plate H (or sometimes h) denotes the separation efficiency of a column; thus, a lower value for H denotes a more efficient column.

It should be pointed out that H and N may also vary somewhat with the retention factors of the analytical sample components. Typically, analytes with low k-values generally have higher plate numbers (sharper, better resolved peaks) than those with higher k-values (4 to 10, for example). It is also well known that H (and therefore N) changes with the linear flow rate (μ) of the eluent or mobile phase. The relationship will be discussed with the rate theory of chromatography below, but H generally decreases with μ.

Another important variable in HPLC is the "peak capacity", which determines the number of peaks that can be present in the chromatogram. Guidelines for peak capacities as a function of the number of theoretical plates can be given for some typical ranges of retention factors. A typical packed column with 5000 plates will yield a peak capacity of about 17 for k-values ranging from 0.2 to 2, and about 50 for k-values ranging from 0.5 to 20. Note that, as the retention factor increases, the peaks become broader, so reducing the peak capacity; however, a wider range of retention factors will yield a system with a larger peak capacity.

As we have stated, a satisfactory chromatographic separation depends on having a column with a sufficient plate number, N, as well as an adequate difference in k-values. Peak resolution (R_s) in terms of the separation factor ($\alpha = k_2/k_1$), the average retention factor, $k_{av} = (k_1 + k_2)/2$, and the plate number, N, is given by:

$$R_s = \frac{\alpha - 1}{\alpha + 1} \cdot \frac{\sqrt{N}}{2} \cdot \frac{k_{av}}{1 + k_{av}} \tag{A1.12}$$

This relationship is often used to estimate the number of plates needed for a separation.

$$\sqrt{N} = \frac{2(\alpha + 1)}{\alpha - 1} \cdot \frac{1 + k_{av}}{k_{av}} \cdot R_s \tag{A1.13}$$

For example, if $k_2 = 4.2$ and $k_1 = 3.8$, $\alpha = 1.08$, for $R_s = 1.0$:

$$\sqrt{N} = 2 \cdot \frac{2.08}{0.08} \cdot \frac{5}{4} \cdot 1.0 = 65$$

$N = 4200$ plates

In reversed-phase liquid chromatography, the separations are based on differences in the partitioning of sample compounds between a rather hydrophobic stationary phase and a mobile liquid phase, such as acetonitrile/water. The retention factors are generally increased as the molecular size and hydrophobicity of the organic molecules become larger. The values of k become smaller as the proportion of organic solvent in the mobile phase is increased. The retention factors of the analytes to be separated are maintained in a desirable range by adjusting the composition of the mobile phase.

In ion-exchange chromatography, the retention factor of an analyte is again determined by its relative affinity for the stationary and mobile phases, but the mechanism is different. Here, the stationary phase is an ion-exchange material, and the sample ions are retained only at specific ionic sites on the ion exchanger. In order for the sample to be attracted to the exchange site, an eluent ion of the same charge (positive or negative) must be displaced. The retention factor of the sample ion is kept within the desired range by adjusting the concentration of the competing ion in the eluent.

A specific example may be used to illustrate these principles. An anion exchanger (Anex) is converted to the Cl^- form by passing a solution of sodium chloride through the ion-exchange column. The introduction of a sample containing RNA PO_4^- analyte sample compounds sets up the exchange equilibrium:

$$xAnex - Cl^- + RNA\ PO_4^- \rightleftarrows Anex - RNA\ PO_4^- + xCl^- \tag{A1.14}$$

where PO_4^- represents the RNA phosphodiester backbone. As each RNA molecule has several phosphate groups, an equal number of Cl^- are displaced on the Anex. By adjusting the Cl^- concentration in the eluent to an appropriate value (1 M in some instances) the ratio of Anex–RNA PO_4^- to RNA PO_4^- in solution can be controlled; hence, the retention factor can be kept within desired limits.

The column packing or resin capacity is generally directly proportional to the retention of the sample compounds. The resin capacity is the amount of stationary phase that can actively interact with the sample compound. In the equations described above, the example was based on an insoluble, immobile liquid that was on the column packing particles. While the early chromatographic studies used particles of this type, modern chromatographic packings are usually solid materials that consist of a solid substrate onto which the functional groups are chemically attached. In the case of reverse-phase materials, the surface is a neutral organic material usually consisting of an alkyl group which is 18 carbons long. The column particle capacity is based on the amount of functional groups that are bonded to the surface of the bead. In some particles, the functional groups are attached to substrate both in the interior of the beads and on the outside. If the sample size is too large to enter the particles, then only those stationary phase sites that are accessible to the sample can be considered as contributing to the capacity of the packing material.

Ion exchangers behave in a similar manner, with the retention increasing with ion-exchange capacity. Ion-exchange capacity is usually denoted as the number of milliequivalents of ion-exchange sites per gram or substrate (mEq.g^{-1}).

A1.3
Source and Effect of Peak Diffusion

A famous equation used by chromatographers to measure the effect of these parameters on peak broadening is the van Deemter equation [4]:

$$H = A + B/\mu + C\mu$$

where H is the height equivalent of a theoretical plate and μ is the linear flow rate of the liquid through the column (expressed in units of mm of column length per second). The linear flow rate is a function of flow rate, column diameter, and the fluid volume of the column. Of course, linear velocity is directly proportional to eluent flow rate, and is inversely proportional to the square of the column radius. The fluid volume (the volume of liquid between the packing material and liquid in the packing pores) of the column depends on how well the particles are packed into the column and the flow through pore volume of the column material. According to the van Deemter equation, there are three principal contributions to the broadening of a peak signified by the A, B, and C terms.

The A Term: This is also called the "multipath" or "eddy diffusion". In a column, the fluid travels many paths at random. However, as some paths are longer than others, some sample molecules will move slowly than others while passing through the column. In other words, the flow velocity through a packed column will vary widely with the flow path through the column, and this will result in peak broadening. Some sample molecules will travel more rapidly by following open pathways (channeling); others will diffuse into restricted areas and lag behind the zone center (eddy diffusion). In a poorly packed column, flow along the column wall can be quite different than flow through the column bed. These differing flow velocities will cause zone dispersion about the average velocity. The use of a homogeneous column packing of regular shape and narrow particle size range will minimize multipath diffusion; the column material should also be packed very evenly into the column. The A-value will normally increase with the average diameter of the packing material. Also, it is important not to leave off the column end so that the column can partially dry, as this may cause shrinkage of the bed and increase the value of A. The A term is independent of the flow rate, but is instead a function of the column packing.

The B Term: The molecular longitudinal diffusion B term represents the zone spreading that each sample component exhibits due to diffusion along the column axis (longitudinal diffusion). As diffusion coefficients in aqueous solution are generally low, the contribution of this term is relatively small, unless the retention time is quite long due to a very slow flow rate or a high retention factor. The contribution of this term to band-broadening depends on the residence time of the sample compound in the column. Therefore, the use of a larger eluent flow rate will minimize this term. However, it is only at extremely low flow rates where the B term becomes significant; hence, it is mostly ignored in liquid chromatography.

The C Term: This term – the resistance to mass transfer – relates to the rate at which a sample compound travels to and from the mobile phase and the stationary phase. Mobile phases that have a high viscosity and a low transfer rate would contribute to peak broadening. There is also an additional rate of interaction of the sample material with the stationary phase. For example, if the stationary phase is a solid nonpolar surface, the kinetic rate of adsorption and desorption kinetic will form part of the chromatographic process. Moreover, if the stationary phase is an ion exchanger, then the kinetic rate of exchange process itself will form part of the chromatographic process. Resistance to mass transfer is by far the major contributor to sample zone spreading within the column. However, the C term can be minimized by using a column packing that attains equilibrium of the sample as quickly as possible between the mobile and stationary phases. In other words, the stationary phase sites are completely accessible to the mobile phase. In many cases, a higher eluent temperature will also help.

The contributions of each of these terms in the van Deemter equation are shown in Figure A1.3, where a plot of H as a function of linear flow rate μ is given. The objective is either to minimize H or to minimize the time required for separation by using the fastest flow rate that will provide a reasonably low value of H. The A term is usually low for a well-packed column, and is not affected by flow rate, but as the flow rate increases the B term will become smaller and the C term larger. As B is divided by μ, in theory this term will go to infinity as μ approaches zero. However, the separation time increases with a decreasing eluent flow rate, and the

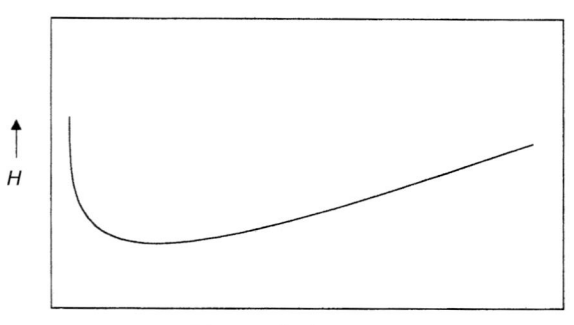

Figure A1.3 The van Deemter plot shows the optimum linear velocity of the eluent to achieve lowest H (height equivalent of a theoretical plate) and the greatest peak resolution. The column length divided by H will give the plate count for a particular column; as H decreases, the total number of theoretical plates increases for the column. At very low linear velocities, the column becomes very efficient, but the separation time also increases. At extremely low linear velocities, the peaks broaden somewhat due to linear diffusion, but there is no advantage to slowing down the separation with extremely low eluent flow rates.

flow rate at which the B term becomes significant is so low that it would lead to impractically long separation times. At some optimum value of μ these terms will balance each other, such that H and the separation time will be at an optimum. Generally, for most packed columns the volumetric flow rate is about 0.5 to 2.0 ml min^{-1}.

A1.4
Extra-Column Effects Causing Peak Broadening

Substantial peak broadening may also occur *outside* the column. To avoid this, the transfer lines connecting the column should be as short as possible. Stagnant areas in the system must also be avoided; these can occur, for example, if the connection between two pieces of tubing is not fitted correctly. On addition, the detector cell should have a low dead volume. These extra-column effects can be measured by injecting a marker (UV-absorbing sample) into the chromatographic system, with the column by-passed by removing it from the instrument and connecting the inlet and outlet tubing together with a union connection. Now, when an injection is made, the sample travels directly from the injector through the tubing into the detector. If the system has a low dead volume, the peak from this injection should be sharp and symmetrical; in that case, the extra-column effects are minimal. However, if the peak from this injection is tailing and not symmetrical, then there is a dead volume in the tubing and connections that prevent the efficient and direct transfer of liquid through the chromatographic system. In this situation, each connection should be examined to ensure proper connection. The tubing should be examined to ensure the diameter following the instrument manufacturer's recommendations.

A system with a low extra-column volume will show a sharp, nontailing peak, whereas a tailing or broad peak indicates a nonoptimized system.

References

1 Poole, C.F. and Poole, S.K. (1991) *Chromatography Today*, Elsevier, New York.
2 Snyder, L.R. and Kirkland, J.J. (1979) *Introduction to Modern Liquid Chromatography*, 2nd edn, John Wiley & Sons, Inc., New York.
3 Heftmann, H. (ed.) (1992) *Chromatography, Part A: Fundamentals and Techniques*, 5th edn, Elsevier, Amsterdam.
4 Ettre, L.S. and Meyer, V.R. (2000) Two symposia, when HPLC was young. LC/GC Magazine, July 2000.
5 Horvath, C. (1974) *75 Years of Chromatography – A Historical Dialogue* (eds L.S. Ettre and A. Zlatkis), Elsevier, Amsterdam, pp. 704–14.
6 Giddings, J.C. (1965) *Dynamics of Chromatography*, Marcel Dekker, New York.
7 Katz, E., Ogan, K.L. and Scott, R.P.W. (1983) Peak dispersion and mobile phase velocity in liquid chromatography: the pertinent relationship for porous silica. *J. Chromatogr.*, **270**, 51.

8 Grushka, E. and Kikta, E. (1975) Chromatographic broadening technique of liquid diffusivity measurements. *J. Phys. Chem.*, **79**, 2199.

9 van Deemter, J.J., Zuiderweg, E.J. and Klinkenberg, A. (1956) Longitutional diffusion and resistance to mass transfer as causes of nonideality in chromatography. *Chem. Eng. Sci.*, **5**, 271.

Appendix 2
HPLC Instrumentation and Operation

A2.1
General Description of the RNA Chromatograph

Figure A2.1 shows a block diagram of the general components of a RNA Chromatography instrument. The components are:

- A supply of eluents or buffers (also called the mobile phase).
- An eluent degasser to remove dissolved oxygen, carbon dioxide and nitrogen from the fluid.
- A high-pressure pump (with pressure indicator) to deliver the eluent or mobile phase.
- An autosampler, including a syringe and sample valve for introducing the sample into the eluent stream and onto the column.
- A column (also called the stationary phase) to separate the sample mixture into the individual components.
- An oven to contain the column and control the temperature of the nucleic acids.
- A detector to measure the peaks or bands of the nucleic acid fragments as they elute from the column.
- An optional fragment collector to collect any particular fragment of interest.
- A data system for collecting and organizing the chromatograms.

A computer is used to control the instrument, and software is used to develop and implement analytical methods for each set of samples. The software guides the user to calibrate the instrument and to develop the appropriate and optimum methods, depending on the analysis being performed and whether fragments are to be collected.

A key component of the instrument is the *separation column*. Unlike electrophoresis gels, which are used only once, the chromatographic column is used for several hundreds (or even thousands) of injections before it is replaced. The eluent is pumped continuously through the column. Periodically, samples are injected and separated. A particular mix of eluent strength is used for each separation depending on the sample type (fragment sizes) and the desired separation.

A *pump* is used to force the eluent or buffer through the system, including the column, at a fixed rate (e.g., $0.9\,\text{ml}\,\text{min}^{-1}$). In the sample FILL mode, a small

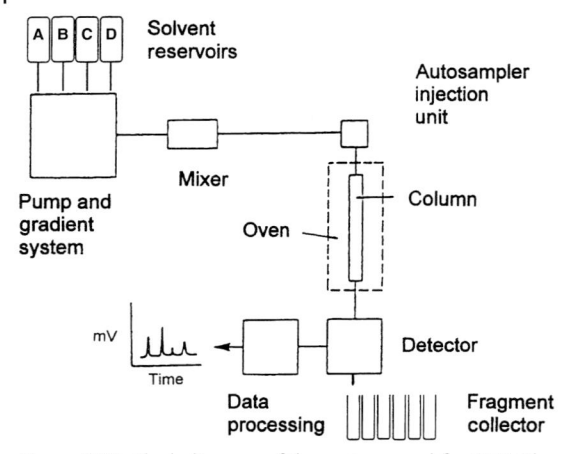

Figure A2.1 Block diagram of the system used for RNA Chromatography.

volume of sample (typically 4 µl) is pulled into the autosampler syringe and placed into the sample loop. At the same time, eluent is being pumped through the rest of the system, while bypassing the sample loop. In the sample INJECT mode, a valve is turned so that the eluent sweeps the sample from the sample loop into the column.

A *detector cell* of low dead volume is placed in the system just after the column. The detector is connected to a data-acquisition device so that a chromatogram of the separation (signal versus time) can be plotted automatically. A UV-visible detector is most often used in RNA Chromatography. Because RNA absorbs UV light, a signal results when the fragment band travels past the detection window. This absorption of light is detected and amplified and recorded as a function of time to produce a chromatogram.

Finally, a *fragment collector* can be positioned at the outlet tube of the detector. When a fragment of interest is detected the deposition probe can be directed to a collection vial. After the material has been collected the probe is directed back to waste.

A2.1.1
General Instrument and Materials

This section provides a detailed description of the various components of a RNA Chromatography instrument, their function, and some general points for upkeep of the chromatograph. New users can use the information not only to operate an instrument, but also to understand how an instrument is built and to recognize the parts of the instrument that may require regular maintenance. The hardware is similar to that used for HPLC, but does have some important differences. Those readers familiar with HPLC will recognize the similarity and the differences to the RNA Chromatograph.

A2.1 General Description of the RNA Chromatograph

The reader should again refer back to Figure A2.1. Fluid is pumped through the system under high pressure. Some of the backpressure produced is due to the small-diameter tubing, but most of the pressure is due to the separation column, which is located in the oven. Everything on the high-pressure side – from the pump outlet to the end of the column – must be strong enough to withstand the pressures involved. The wetted parts are usually made from stainless steel, titanium, polyetheretherketone (PEEK) and other types of plastic. Materials, such as sapphire, ruby, or even ceramics are used in the pump heads, check valves, and injectors of the system. PEEK and titanium are the materials of choice for the RNA Chromatograph. Stainless steel is also excellent, provided that the system is properly conditioned to remove and control internal corrosion. In fact, stainless steel components are considered to be more reliable than those made from plastics, but require more care. A stainless steel chromatograph will normally be delivered from the manufacturer pretreated so that corrosion is not present. The reader is advised to consult the instrument manufacturer for care and upkeep instructions. More discussion on the effect of instrument corrosion is included in Appendix A3.

A2.1.1
Dead Volume

The dead volume is any empty space or unoccupied volume. The dead volume of a chromatographic system is between the point where the sample is introduced (the injector) and the point where the peak is detected (the detection cell), and must be kept to a minimum. The presence of too large a dead volume can lead to extreme losses in separation efficiency, due to broadening of the peaks. Although all regions in the flow path are important, the most important region where peak broadening can occur is in the tubing and connections from the top end of the separation column to the detector cell. It is important to follow the manufacturer's instructions when changing the column for making other fluid connections. If fragments are to be collected, the dead volume from the end of the detector cell to the end of the deposition probe is also extremely important for maintaining sharp and crisp fragment collections with no cross-contamination of neighboring peaks.

Of course, there will always be a natural amount of dead volume in a system due to the internal volume of the connecting tubing, the interstitial spaces between the column packing beads, and so on. However, the use of small-bore tubing (0.18 mm, 0.007 inch or smaller) in short lengths when making the injection-to-column and column-to-detector connections is important. It is also important to make sure that the tubing end does not leave a space in the fitting when the connections are made. In general, the tubing should be butted against the bottom of the fitting first, and the screws of the fitting then tightened (see Figure A2.2).

The dead volume before the pump, and from the pump to the injector, should also be reduced to facilitate rapid changes in the eluent composition in gradient elution. A certain amount of dead volume is needed to ensure complete mixing of the different eluent components before they reach the column. However, too much dead volume will lengthen the time needed to change the eluent concentra-

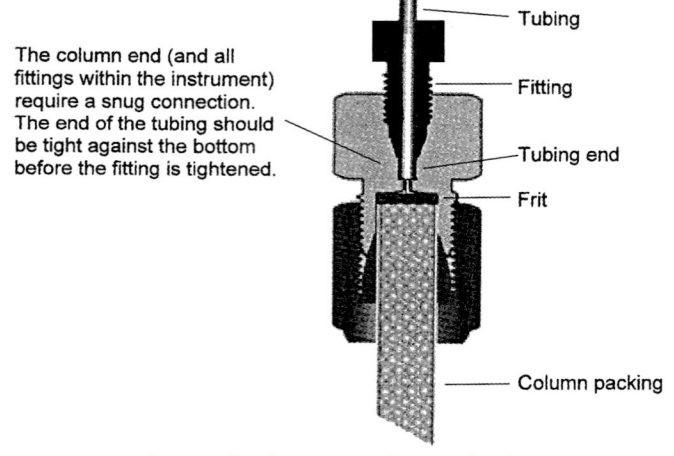

Figure A2.2 Column end and connection fitting with tubing. Note that the dead volume is minimized by ensuring that the end of the tubing is butted down against the fitting.

tion that is part of the gradient process, thus leading to longer separation times. Gradient formation is discussed later in Section A2.1.5.

Eluent entering the pump – and, even more importantly, entering the column – should not contain any dust or other particulate matter, as this can interfere with the pumping action and damage the seal or valves. Material can also collect on the inlet frits or on the inlet of the column; this will cause pressure build-up and lead to premature column replacement. The components used to prepare the eluents are normally filtered, using a 0.2 or 0.45 µm nylon filter. Some filters are made from other materials, and the user should determine that the filter material is compatible with the solvent before use. It should be noted that, after the eluent has been prepared it is normally not filtered because the acetonitrile solvent is quite volatile, and filtering may remove it or change its concentration in the eluent. This would in turn alter the eluting strength of the eluent. For this reason, glassware and other containers used to prepare and store the eluent should be rinsed with particulate-free solvent and dried carefully before use. (Again, there should be no residual solvent in the containers that would alter the concentration of the eluent.)

A2.1.3
Degassing the Eluent

Degassing the eluent is important because air can become trapped in the pump check valves (see Section A2.1.4) and cause the pump to lose its prime. Loss of prime results in an erratic eluent flow, or no flow at all. Occasionally, only one pump head will lose its prime and the pressure will fluctuate in rhythm with the pump stroke. Another reason for removing dissolved air from the eluent is because oxygen in the fluid can cause it to become highly corrosive. Removal of the oxygen through degassing will result in less required maintenance of the stainless steel components of the chromatograph.

Usually, degassed water is used to prepare eluents and every effort should be made to ensure that the exposure of eluent to air is kept to a minimum after preparation. The best way to remove these gases is to use an inline degasser, and these are becoming quite popular. They are small devices that contain two to four channels through which the eluent travels from the reservoir to the pump. The tubing in the device is gas permeable and surrounded by vacuum; this causes gases in the fluid to be transported through the tubing wall to the vacuum, leaving the eluent ions and higher-boiling fluids behind.

It should be noted that when the pump flow is stopped or is on standby, this will cause the eluent to have a very long residence time in the degasser. Under these conditions, the organic solvent (acetonitrile) in the eluent will gradually pass through the membrane and alter the concentration of the eluent; therefore, it is best to flush this fluid out of the instrument when it is first started up. The best method to accomplish this is to perform a "blank gradient" run, where no sample is injected but the pump goes through the gradient process. Failure to do this would result in the first real run to be erratic. Finally, it is best to change the eluents every day (or at least every couple of days) in order to keep the concentration accurate, and also to prevent bacterial growth in the reservoirs. If bacterial material is pumped through the system it is likely to build up at the top of the column and increase the backpressure of the system, sometimes quickly and dramatically, and at other times only gradually.

A2.1.4
Pumps

Chromatographic pumps are designed around an eccentric cam that is connected to a piston (Figure A2.3). The rotation of the motor is transferred into the

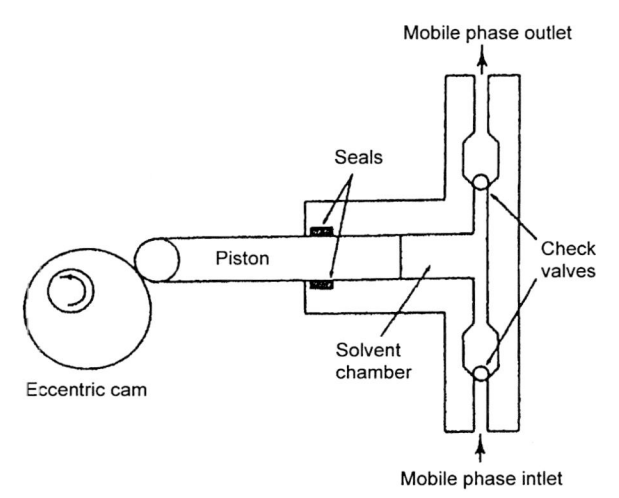

Figure A2.3 The cam, pump head, piston, piston seals and check valves of the RNA chromatograph.

reciprocal movement of the piston. A pair of check valves controls the direction of flow through the pump head (discussed below). A pump seal surrounding the piston body keeps the eluent from leaking out of the pump head. A piston seal leak is usually formed at the reverse side of the pump head.

In single-headed reciprocating pumps, the eluent is delivered to the column for only half of the pumping cycle. A pulse dampener is used to soften the pulsating action of pressure of the pumping cycle and to provide an eluent flow when the pump is refilling. The use of a dual-head pump is better because the heads are operated 180° out of phase with each other–one pump head is pumping while the other is filling, and *vice versa*. The eluent flow rate is usually controlled by the pump motor speed, although some pumps are available that control the flow rate by control of the piston stroke distance.

Figure A2.4 shows how the check valve works. On the intake stroke, the piston is withdrawn into the pump head, causing a suction that forces the outlet check valve ball to settle onto its seat and prevent back flow. At the same time, the inlet check valve ball rises from its seat, allowing eluent to fill the pump head. The piston travels then back into the pump head on the delivery stroke. The pressure increase of the eluent causes the inlet check valve ball to seat, thus preventing back flow, and at the same time the outlet valve is opened. The eluent is forced out of the pump head through the outlet check valve, through the autosampler injection valve, and into the column. Failures of either of the check valve balls to seat properly will cause pump head failure so that the eluent will not be pumped. In most cases, this is due to air being trapped in the check valve so that the ball cannot seat properly. Flushing or purging the head usually will take care of this problem. The use of degassed eluents is also helpful. In a few cases, particulate material can prevent seating of the check valve, and the check valve will need to be cleaned by flushing, or replaced. Dilute nitric acid is often used as a cleaning solvent in this process. The instrument manufacturer will provide instructions on how to perform this operation.

A2.1.5
Gradient Formation

Isocratic separations are performed using an eluent at a constant or uniform concentration of eluent solution. While it is desirable (simpler) to perform nucleic acid separations with single isocratic eluent, it is almost always necessary to perform them with an eluent gradient to achieve the desired peak resolution. A weak eluent is first pumped through the column as the sample is injected. The eluent strength is then increased gradually over the course of a chromatographic run, allowing the separation of nucleic acids that may have a wide range of affinities for the column. Weakly adherent nucleic acid fragments will elute first, but as the eluent concentration is increased more strongly adherent fragments will be eluted.

There is the old adage that a weakness of something can become its strength if only used in the correct way. When compared to electrophoresis, where gradients

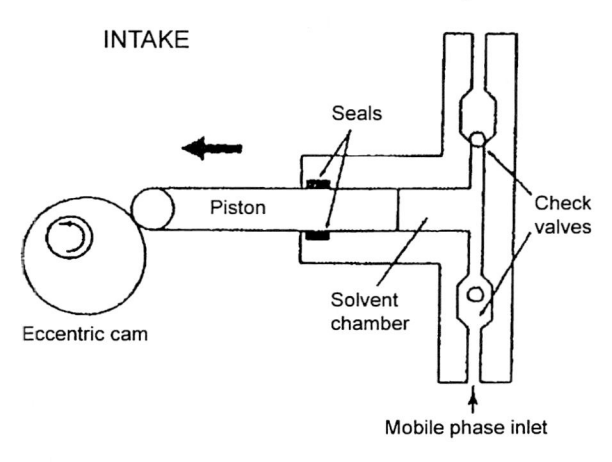

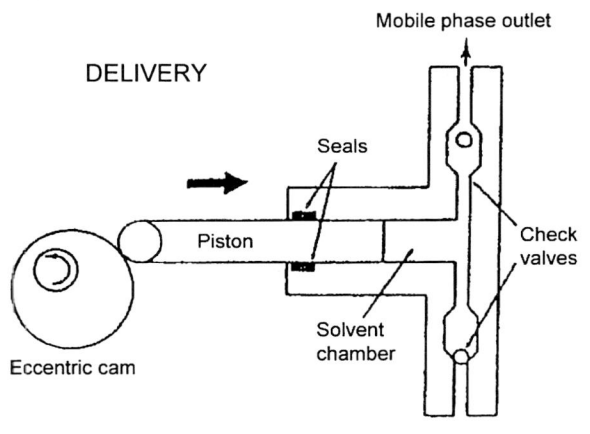

A2.1 General Description of the RNA Chromatograph

Figure A2.4 Check valve positions during intake and delivery strokes of the pump head piston.

are rarely used, gradients in RNA Chromatography add a complexity for fragment separation. But this is also one of the major strengths of RNA Chromatography. The concentration of acetonitrile present in the column at any particular instant dictates the size of the fragment that is being separated by the column. The separation conditions can be controlled either in "real time" or instantly, so that targeted materials can be eluted and collected as desired. Smaller fragments require a lower concentration of acetonitrile, and larger fragments a higher concentration. The range and the duration of the gradient will determine how fast the separation will be, and what fragment size range will be measured, and this feature can be changed from one separation run to the next at will.

High pressure mixing gradient system

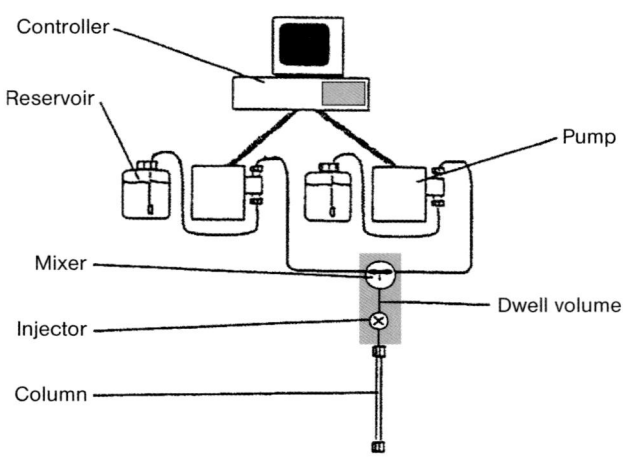

Figure A2.5 High-pressure mixing systems use two or more independent pumps to generate the gradient. The advantages of high-pressure mixing are a smaller dwell volume and faster gradient formation.

Figures A2.5 and A2.6 show two most popular methods for forming gradients. In *high-pressure gradient mixing*, the flows from two high-pressure pumps—one flow containing a weak eluent and the other a stronger eluent—are directed into a high-pressure mixing chamber. The mixed flow is directed to the injector and then onto the column. By controlling the relative pumping rate delivery of each pump it is possible to form a gradient. If the total combined flow from the two pumps is held constant, a high flow from the weak eluent pump and a low flow from the strong eluent pump will initiate the gradient run. Over the course of the chromatographic run the relative flow rate of the strong eluent pump is increased, while that of the weak eluent pump is decreased, keeping the total flow rate constant.

A more popular and rugged method of forming gradients is that of *low-pressure gradient mixing*, where a single pump contains three or four microproportioning valves at the inlet of the pump. At low pressure, gradients can be formed from eluents A, B, C, and D (or any combination) by metering controlled amounts from the various eluent reservoirs into the pump. The composition in the low-pressure mixing chamber is controlled by using timed proportioning valves. Only one reservoir valve is open at any time. The total cycle time remains constant throughout the gradient, but the time that any one of the reservoir valves remains on will vary. At the start of the gradient, the valve connected to the weak eluent is open longest, but as the gradient progresses the valve connected to the strong eluent will be opened for increasingly longer periods, and the weaker eluent valve for shorter periods. The time cycle of the valve remains constant, however. Although, generally the gradients are formed with just two valves, on occasion up to four valves

A2.1 General Description of the RNA Chromatograph | 167

Low pressure mixing gradient system

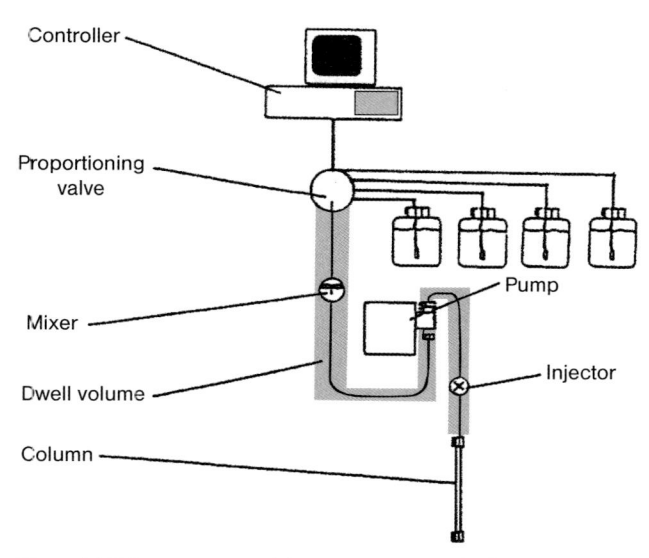

Figure A2.6 Low-pressure mixing systems use a single pump with a proportioning valve to control the composition. The advantages of low-pressure mixing are a lower cost (single pump) and more versatile gradients (four solvents).

may be available to provide options for different types of gradient, or for the use of cleaning solutions.

This method of gradient generation is less expensive than high-pressure gradient formation because only one pump is used. It is also usually more rugged because the extra pump and check valves are more likely to fail than is the microswitching valve of a low-pressure gradient system. On the other hand, high-pressure gradients can be faster because there is less dwell volume (gradient delay volume) in this gradient formation system, and this results in the gradient reaching the column and detector more quickly. The actual time of gradient formation depends in part on the volume of the mixer located after the eluent gradient has been first formed. As the mixers for both types of gradient tend to have large volumes (to ensure complete mixing), the time for the gradient to reach the column is usually not much different for the two types of system.

A2.1.6
Pressures

Column inlet pressures can vary from 500 p.s.i. up to perhaps a high of 3500 p.s.i., with normal operating pressures around 1500 p.s.i. The pressure limit on a RNA chromatograph is usually 4000–5000 p.s.i., depending on the fittings and other hardware used. The eluent backpressure is directly proportional to the eluent flow rate. Although still popular, p.s.i. (pounds per square inch) is gradually being replaced by more modern terms for pressure measurement; that is 1 bar = 1 atm (atmosphere) = 14.5 p.s.i. = 10^5 Pa (Pascal).

A2.1.7
Autosampler Injector

The injections are normally taken directly from a 96-well or 384-well PCR thermal cycler plate. Although it is possible to inject a sample manually, this is rarely done. In any case, both manual and autosamplers rely on the *injection valve*, which is designed to introduce precise amounts of sample into the sample stream. The variation of sample volume injected is usually less than 0.5% from one injection to the next. Figure A2.7 shows a schematic of the valve, which is a six-port, two-position device, where one position is "load" and the other is "inject". In the load position, the sample is first cleaned with an external solution, after which the syringe takes up the sample from the PCR vial and pushes it into the injection loop. (Note that the external syringe needle cleaning solution must contain an organic solvent concentration that is less than or equal to the starting gradient, as this will prevent premature initiation of the gradient.)

The loop may be either partially filled (partial loop injection) or completely filled (full loop injection) (see Figure A2.7). *Partial loop injection* is by far the most popular approach, due to the small volume injections performed in RNA Chromatography. In this procedure, most or all of the sample taken up by the syringe is ultimately injected into the system. Partial loop injection depends on a precise filling of the loop with a small known amount of material; the loop must not be filled to more than 50% of its total volume, or the injection precision may be lost. In *full loop injection*, the sample is pushed completely through the loop, the typical

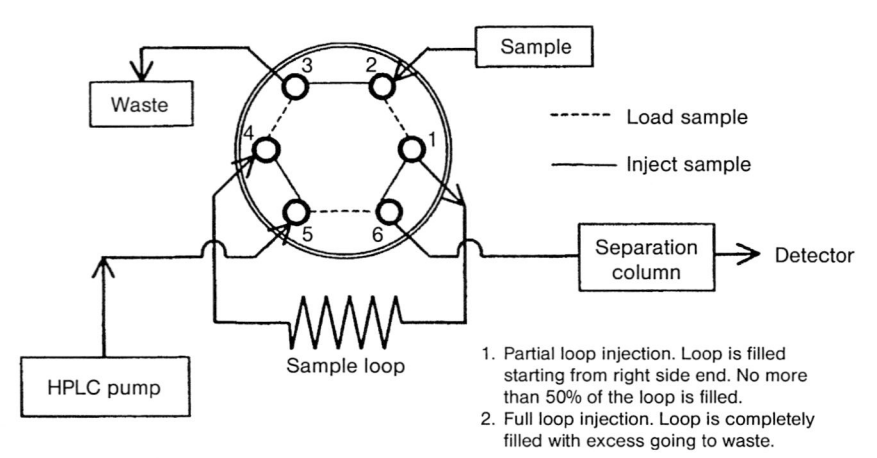

Figure A2.7 Schematic representation of the injection valve, as found in the autosampler, depicting the partial and full loop injection methods.

sizes of which range from 10 to 200 μl. Normally, at least a twofold amount of sample is used to fill the loop, with excess sample from the loop going to waste.

While the loop is being loaded with sample, the eluent is traveling in the bypass channel of the injection valve to the column. An injection of the sample is accomplished by turning the valve and placing the injection loop into the eluent stream. Usually, the flow of the eluent is opposite to the flow of loading sample into the loop. The injected sample travels to the head of the column as a bolus of fluid; the fragments in the sample subsequently interact with or are adsorbed onto the column, and the separation process is then started with an eluent of appropriate strength pushing the sample components down the column. Injection valves require periodic maintenance, and normally need to be serviced after about 5000 injections. The manual for the instrument should be consulted for all details on service.

A2.1.8
Separation Column

The separation column (sometimes called a "cartridge") is a small cylinder that contains the packing material used to separate the nucleic acid fragments. The packing materials are held inside the column by porous "frits" which are located at each end of the column and allow the fluids to enter and leave the column.

Columns for chromatography are packed with beads that provide the basis for the separations. The column packing is normally performed by capping the bottom of the column with a frit and connecting the column to a reservoir. A slurry of the packing material and a solvent is prepared and poured into the reservoir; a pump is then connected to the top of the reservoir and the fluid pumped under high pressure, forcing the slurry into the column until, eventually, the column is filled. After a suitable time, the pump is stopped, the reservoir is disconnected, and the top of the column is capped. The column is tested before use to ensure that the packing procedure has been successful.

Essentially two types of column material are used for RNA separations. The most popular is a reverse-phase (neutral-charged surface) material; the popular trademark for this is DNASep® (supplied by Transgenomic, Inc., San Jose, CA and Omaha, NE, USA). Another, older type of column material is an anion exchanger. Although, many recent studies have been conducted with neutral reverse-phase materials, anion exchangers are still used today, especially for the separation of short, single-stranded DNA.

The most popular column materials are polymer-based because they are rugged and can withstand extremes in eluent pH, although silica-based materials are also available on which a neutral surface has been applied. The column is carefully packed with a spherical anion exchanger with a typical particle diameter of approximately 2 μm. Most column packings are functionalized with a hydrophobic, neutral C18 alkyl group, or with quaternary ammonium groups, which serve as the sites for the anion-exchange process.

A typical column used in ion chromatography might be 50 mm long and have a bed diameter of 4.6 mm, although columns much shorter (e.g., 30 mm) or longer

(100 mm) can be used. For purifying larger amounts of RNA fragments, preparative columns can be used which have larger bed diameters of 7.8 or 11 mm, or even larger.

Reusable PEEK fittings are used almost exclusively to connect tubing to columns and other instrument components. As noted above, the tubing should be bottomed out or pushed completely into the column end before the fitting is tightened, in order to ensure that there is no unnecessary dead volume in the connection.

A2.1.9
Column Protection

Correct column protection will not only extend the useful life of the separation column, but may also result in more reliable analytical results over its lifetime.

Scavenger Columns   These columns, located between the pump and injector, are one means of protecting the column. The scavenger removes particulate material that may be present in the eluent, and may also contain a resin to "polish" the eluent of any dissolved contaminant. An example is a chelating resin for the removal of metal ion or colloidal contaminants. Besides protecting the separation column, scavenger columns may also improve detection of the fragments by reducing the detector background signal due to residual contaminants.

Guard Columns   Another method of column protection is to use a "guard column" or "guard cartridge". The guard is located directly in front of the column, and contains the same or similar packing as the main separation column. The frits located at the ends of the separation column serve as an efficient means to trap particulate materials, including denatured proteins, colloidal metals, and dust. Dissolved contaminants include materials in the sample that may adsorb to the column packing and are not easily removed by cleaning procedures. Hence, particulate or dissolved contaminants that would normally be trapped by the column are instead trapped on the guard column which, being less expensive than the main separation column, can be replaced more often. The main disadvantage of the guard is that it can affect the selectivity or retention times of the separation.

In-Line Filters   Most RNA Chromatography users do not use scavenger columns or guards, but rather prefer the use of in-line filters located directly in front of the separation column. Consequently, material that would normally be trapped and contaminate the separation column is instead trapped by the filter. In-line filters are relatively inexpensive and can (and should) be changed frequently – at least every 2–3 weeks and sometimes much more often. In-line filters do not affect the retention of nucleic acid fragments.

Flow Reversal   Another less common (but still useful) approach for extending the life of the separation column is to reverse the direction of flow through the column. When the column is reversed, any particulate or adsorbed materials that have accumulated at the top of the column in effect become materials at the bottom of the column, and can be washed off with normal operation of the system. Reversal of the column can be done every 2–4 weeks over the column's lifetime.

Cleaning Eluents   Finally, the column should be cleaned with a *cleaning eluent* throughout its lifetime. The cleaning procedures must be implemented at the end of each run, or after a set of injections. The reason why cleaning is important is that the concentration of the (acetonitrile) solvent used in the eluent during normal operation is quite low (rarely above 25%). While this provides a useful separation of nucleic acids, any contaminants usually require much higher concentrations of acetonitrile to be washed from the neutral column packing surface. The instrument manufacturers take this into account by providing a cleaning solvent for use at the end of each run, or when a set of runs has been completed. Failure to keep the column clean is probably the most common error that the user can make. For this reason, standards with known peak patterns should be injected at the beginning and end of a series of runs to ensure that the column is clean and working properly throughout the analysis of the samples.

A2.1.10
Column Oven

The column oven forms a valuable part of any RNA Chromatography instrument [1]. One benefit of using a column oven is the improved detection of RNA fragments. UV detection can be affected by changes in the refractive index of the fluid entering the detection cell, and these changes in turn contribute to background detector signal noise. Although the background noise caused by refractive index changes is usually small, it can still affect the detection limits when very low concentrations of materials are being measured. Controlling the temperature of the fluid entering the detection cell can minimize the noise due to refractive index changes.

However, the most compelling reason to use a column oven is to control the structure of the RNA. By controlling the temperature of the oven, the separation conditions can be characterized for double-stranded (nondenaturing), secondary structure (partial denaturing), and single-stranded (full denaturing) nucleic acids. Full denaturing conditions will control (or at least modify) any secondary structure that might be present in single-stranded RNA nucleic acids, and may also affect the retention of nucleic acids, resulting in either multiple or changed retention.

Temperature should be thought of as a reagent in RNA Chromatography. Virtually all ovens that are used for normal HPLC operation are unsuitable for RNA Chromatography, because it is not the control of the oven that is important, but

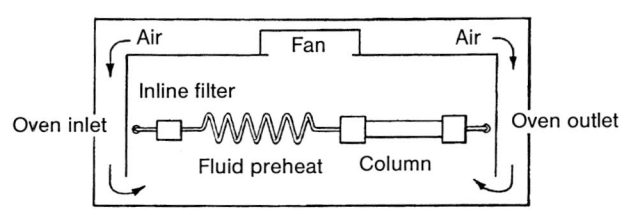

Figure A2.8 Schematic representation of the oven compartment, showing the preheat coil, the inline filter and placement of the separation column. The eluent must be heated up to the oven set point before entering the column. (Reproduced with permission from Ref. [1]).

rather the control of the eluent fluid entering the column. Because the fluid entering the oven compartment is cooler than the oven, there will be a time lag before it reaches oven temperature (in most HPLC ovens, this never occurs). To overcome this problem, a preheat coil is positioned ahead of the column (see Figure A2.8). As the fluid passes through this coil, the eluent (and injected sample) are gradually heated so that, when they enter the column, they have been heated to an accurate temperature. The oven temperature should also not be allowed to drift; in fact, it should be very precise, always reaching the same specific temperature as directed by the control system.

As the oven is one of the most critical and difficult parameters to control in RNA Chromatography, it is advised that the operator should consult the manufacturer for information on its use, calibration, and upkeep.

A2.2
Detection

A2.2.1
Selective versus General Detection

As the name implies, *general detectors*–of which the UV detector is an excellent example–can be used to detect all nucleic acids.

However, *selective detectors* will detect only those fragments that have a certain detectable property; an example is that of fluorescence, where only those fragments tagged with a fluorescent molecule will be detected. (Note: an exception to this, where fluorescence detection can be converted to general detection, is discussed later.) A selective detector can be extremely effective in picking out a single fragment of interest from either a high eluent background or a high sample matrix background. Another key advantage to selective detection is the possibility of achieving lower detection limits. A selective detector may, or may not, have a higher sensitivity (signal per unit concentration) for a particular ion. However, the lower background signals produced by selective detection will translate into lower detection limits because the signal-to-noise ratio (SNR) is improved. At this point it is important to review some definitions:

- *Detector sensitivity* is the amount of signal that is recorded per unit of sample concentration.

- The *detection limit* depends on the sensitivity and the detector noise. The detection limit as defined by analytical chemists is the amount of sample detected at a signal that is threefold the detector noise level. However, practical detection limits exist when the signal is 10-fold the detector noise level.

One of the great strengths of RNA Chromatography over capillary and slab gel electrophoresis is that UV detection can be used without any previous modification of the nucleic acid. Neither is any tagging is necessary for the separation and purification of nucleic acids.

A2.2.2
Ultraviolet-Visible (UV-VIS) Detectors

As the detection of nucleic acids can be achieved using UV spectrophotometry, it is advisable that the reader have a basic understanding of how a UV detector operates, in order obtain the best performance from these systems (which may also be called UV-VIS detectors).

Nucleic acids absorb light strongly in the UV region, with a maximum wavelength at 260 nm. Variable-wavelength detectors are set at 260 nm for detection, while single-wavelength detectors function very well at 254 nm. In this case, an eluent has been selected *not* to absorb at these wavelengths. However, in some cases where there is absorption of the eluent due to contamination or decomposition, the baseline stability may be poor or the baseline will shift when a gradient run is performed. Whilst a natural baseline shift will always occur when a gradient is performed, too-great a shift or an unreproducible baseline will render the instrument unusable.

The detector must be able to "pick out or see" nucleic fragments in the presence of various components of the eluent molecules and solvent. Peaks containing as little as 0.3 ng of material are detectable, although normally it would be preferable to work with amounts 10-fold higher. A UV detector will respond to all nucleic acid fragments passing through the detector cell, whether they are single-stranded RNA, various analogues, or double-stranded RNA.

Figure A2.9 shows, schematically, how the cell detects the RNA fragments. Light is directed through a window in a "Z" cell, while a transducer measuring the light is positioned on the other end of the cell. The light that reaches the detector is measured continuously as a function of time; this measurement is shown as the baseline in chromatographic separations. As the fragments travel through the cell they absorb light, so that less light reaches the detector transducer. This reduction is amplified electronically and results in the "peaks" of a chromatogram. As the amount of fragment material is increased, the amount of light reaching the detector transducer is reduced, thus forming larger or higher chromatographic peaks.

The fundamental law under which UV-VIS detectors operate is the Lambert–Beer law, which can be stated in the following form:

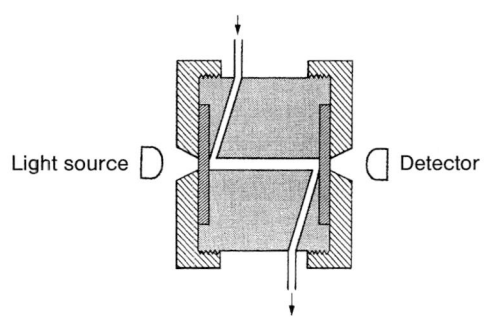

Figure A2.9 Schematic of the "Z" flow cell for a UV detector.

$$A = \varepsilon bC \tag{A2.1}$$

where A is the absorbance of a species of concentration, C, that has an molar absorptivity, ε, in a cell of length b. The concentration is usually expressed in molar terms, and the path length in centimeters. The term, ε, has units that are the inverse of C and b. This leaves A dimensionless; it is usually described in terms of absorbance units (AU). A detector set to a certain sensitivity (e.g., 0.16) is said to be set at 0.16 AU full scale (AUFS) sensitivity. The AU and AUFS may also be expressed as millivolts (mV), or mVFS. The eluent should have a low or zero absorptivity, but if it has any signal at all then the detector must be "zeroed" with eluent flowing through the cell, so that is defined as having no absorbance or no signal.

Although detectors that operate at a single 254 nm wavelength are available, most manufacturers provide UV detectors that have variable-wavelength capabilities, such that any UV wavelength can be chosen for detection. The detectors may also have a visible (VIS) wavelength capability. Typically, wavelengths ranging from 190 to 320 nm are considered UV, while the visible range is from 320 to about 680 nm. The visible wavelength region is rarely used and is unnecessary for RNA Chromatography. Likewise, it is unnecessary to use a variable-wavelength detector, since virtually all measurements are recorded at 260 nm (for single-wavelength detectors, all measurements are at 254 nm).

A2.2.3
Fluorescence Detectors

Fluorescence detection is closely related to absorption detection. When a molecule has absorbed radiant energy and become excited to a higher energy state, it must lose excess energy in order to return to the normal ground electronic state. In this situation, most molecules will simply heat up and shed the excess energy through heat radiation. However, molecules that are rather planar and rigid, and have conjugated double bonds, show a greater tendency to fluoresce; these are most often referred to as *fluorescent dyes*. The most commonly used dyes in molecular biology include FAM, TET, HEX, TAMRA, NED, Pacific Blue, and many others (see Table A2.1).

Fluorescence is the immediate emission of light from a molecule after it has absorbed radiation. With the absorption of the appropriate radiant energy, a molecule is raised from a vibrational level in the ground state to one of many possible vibrational levels in one of the excited electronic levels. The emission of light results from a relaxation to a lower electronic energy state from its excited higher electronic state, producing a wavelength of light corresponding to the decrease in energy state. The absorption and emission of light are specific for that particular fluorescent dye molecule. Typical absorption and emission spectra are shown in Figure A2.10. Here, the excitation spectrum is represented by a plot of different excitation wavelengths versus the emission signal at the optimum wavelength. The emission spectrum is measured by keeping the excitation wavelength constant and measuring the emission signal intensity at various wavelengths. The absorption

Table A2.1 Fluorescent dyes for tagging nucleic acid fragments, and the conditions for excitation and detection. Data assembled from Refs [2, 3]; see also the web sites of www.SyntheticGenetics.com, www.MolecularProbes.com and www.idtRNA.com.

Dye	Excitation maximum (nm)	Emission maximum (nm)
6-FAM (6-carboxy-fluorescein)	494	518
JOE	520	548
TAMRA (carboxytetramethyl rhodamine)	565	580
ROX (Rhodamine X)	585	605
Pacific Blue	416	451
Fluorescein	492	520
HEX (hexachlorofluorescein)	535	556
TET (tetrachlorofluorescein)	521	536
Texas Red	596	615
Cy3	550	570
Cy5	649	670
Cascade Blue	396	410
Marina Blue	362	459
Oregon Green 500	499	519
Oregon Green 514	506	526
Oregon Green 488	495	521
Oregon Green 488-X	494	517
Rhodamine Green	504	532
Rodol Green	496	523
Rhodamine Green-X	503	528
Rhodamine Red-X	560	580
BODIPY FL	502	510
BODIPY 530/550	534	551
BODIPY 493/503	500	509
BODIPY 558/569	559	568
BODIPY 564/570	563	569
BODIPY 576/589	575	588
BODIPY 581/591	581	591
BODIPY FL-X	504	510
BODIPY TR-X	588	616
BODIPY TMR	544	570
BODIPY R6G	528	547
BODIPY R6G-X	529	547
BODIPY 630/650-X	625	640

Data assembled from Refs [2, 3]; see also the web sites of www.SyntheticGenetics.com, www.MolecularProbes.com and www.idtRNA.com.

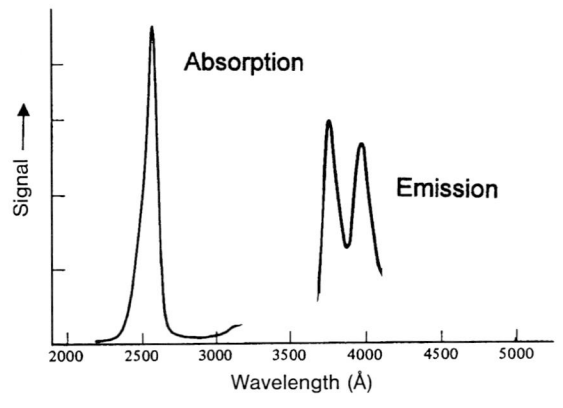

Figure A2.10 Typical absorption and emission spectra for a fluorescent labeling dye. The emission is at a lower energy and longer wavelength than the absorption.

maximum is always at a shorter wavelength (higher energy) than the emission maximum (lower energy) because some energy is lost as heat.

The fluorescence process is rapid (in the order of 10^{-8} s after absorption), and the signal rate high. In most cases, the background fluorescence signal is low, producing a high signal-to-noise level (high detector sensitivity). Normally, a "tagged" or "labeled" nucleic acid molecule will have only one fluorescent molecule attached to each RNA fragment. The labeling process is usually performed as part of a PCR where labeled primers are used to amplify the fragment. Therefore, on a molar basis, the fluorescence is uniform for different fragments; fragments of different sizes will fluoresce to the same extent because they all have one tag per fragment. On a mass basis, the amount of fluorescence per unit mass is dependent on the fragment size; a larger fragment will fluoresce less than a smaller fragment. This is different from UV detection, where sensitivity is based on mass because the entire molecule absorbs light for detection. Even with only one tag per nucleic acid fragment, fluorescence detectability is 30- to 100-fold greater than for UV [2, 3]. Stated in a different way, the detection limits are 30- to 100-fold lower.

Another method of fluorescence tagging has also been used where fluorescent intercalating or grooving reagents are added to the sample prior to injection. In this method, all sequences of the fragment are reacted and the detection is uniform (this type of tagging process will be discussed later).

Figure A2.11 shows, schematically, the arrangement in a fluorescence detector cell. Although several configurations are possible, for conventional detectors the excitation light source is positioned at 90° from the measurement. Stray light is harmful and should be avoided because this will contribute to the background signal. The detector should have variable excitation and variable emission wavelength capabilities, with a diffraction grating normally being used to vary the wavelength. If the type of dye is not varied widely, then optical filters may be used to control the wavelength. The excitation and emission wavelengths of the various dyes are listed in Table A2.1.

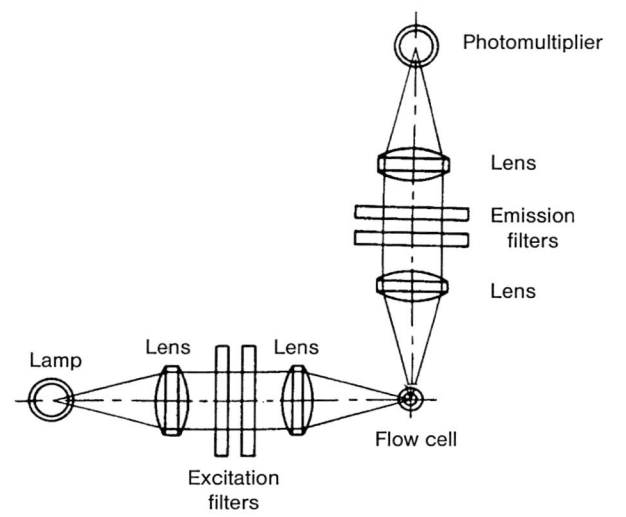

Figure A2.11 Schematic of the cell used for a fluorescence detector.

A2.2.4
Mass Spectrometry Detection

The mass spectrometer is a powerful analytical tool capable of extracting a wealth of information about the structure of organic compounds and complex organic mixtures. In operation, the mass spectrometer produces charged particles that consist of the parent ion and smaller ionic fragments; these are then sorted and recorded according to the mass/charge ratio of each particle. Consequently, every compound will have a characteristic mass spectrum that indicates the different types and relative numbers of each ion.

The mass spectrometric parameters and conditions will affect the type of mass spectrum produced for any particular compound. A detailed interpretation of a sample's mass spectrum permits the determination of a compound's molecular weight, and perhaps also to place functional groups into certain areas of the molecule to determine how it is constructed. Notably, the mass spectrometer can also be used directly on samples, without prior separation by chromatography – an approach which is particularly effective when samples contain only a single compound or limited numbers of compounds of interest. However, if the sample contains several compounds, or if the sample matrix is complex, there are significant advantages to connecting the mass spectrometer to a HPLC system.

A mass spectrometer consists of four parts: (1) the inlet system; (2) the ion source; (3) the analyzer system; and (4) the detector and readout system. A high vacuum is maintained throughout the instrument, from the inlet to the detector, such that a molecular flow of sample fragments to the detector can only be achieved by the (almost complete) removal of gases (air) and solvents. Although connecting a mass spectrometer to a liquid chromatograph provides a powerful analytical combination, it is important to appreciate that, during the procedure the liquid chromatograph will introduce solvent and other materials into the mass spectrometer, placing major constraints on the mass spectrometer inlet, ionization and analyzer methods used for RNA Chromatography. As a consequence, it is necessary to reduce very quickly the pressure experienced by the sample. As the amount of liquid that can be introduced into the mass spectrometer while maintaining an adequate vacuum is limited, the normal method is to use a "splitter", so that only part of the HPLC effluent is injected. In another method, microbore columns that require much lower flow rates for operation can be used; typically, a 1 mm-bore column requires only 5% of the flow rate required by a 4.6 mm-bore column. In another system, a series of inlets is arranged where the vaporized liquid is drawn into a chamber by an auxiliary mechanical pump. The low pressure in the sample inlet reservoir then causes the liquid from the HPLC to be drawn in and vaporized instantly. A heated inlet system will extend the range of use for mass spectrometry (MS) to polar materials and high-boiling compounds that are prone to adsorb onto the walls of the chambers. Unfortunately, however, the temperature that can be used is limited by the thermal degradation properties of the compound; above 200 °C, most compounds containing oxygen or nitrogen will be thermally decomposed.

Past attempts at applying MS to RNA research have proved difficult due to the problems associated with vaporizing and ionizing molecules as large as typical RNA fragments. Nucleic acids have extremely high molecular weights and are nonvolatile. The ionization efficiency of a MS source must be sufficiently high that a large part of the neutral sample particles present in the inlet will become ions. For this, several ionization methods can be used, including electrospray ionization (ESI), matrix-assisted laser desorption, electron impact, chemical ionization, fast atom bombardment, chemical ionization, spark source, and so on. ESI (see Table A2.2) is a gentle or soft ionization method in which the droplets are exposed to a charged particle beam as they are evaporated. The majority of the fragments remain intact when the ions are formed; these fragment ions may be either positive or negative, depending on the mode being used, and will normally have between one and three charges. Nucleic acids are naturally negatively charged due to their phosphate groups; the latter are paired with positive counterions present in the chromatographic effluent, so that the charge associated with the fragment is only that introduced by the electrospray method.

Many different types of mass analyzer are available, depending on the mechanism used to differentiate among charged ions (see Table A2.3). Although the details of these mass spectrometers are beyond the scope of this book, Willard et al. have produced an excellent text on the subject [4]. The mass analyzer that is used most often for nucleic acid chromatography is the "quadrupole", where the field is formed by four electrically conducting, parallel rods that are oriented symmetrically around an ion path. Equal, but opposite, potentials are applied to two pairs of rods, with each potential having direct current (DC) and radiofrequency (RF) components. An ion injected down the axis will travel the length of the rods, without striking any of them, only if it has a charge to mass ratio that corresponds to frequency applied. Altering these electrical parameters will in turn change the

Table A2.2 Types of mass spectrometry ionization methods.

Method	Typical analytes	Sample introduction	Maximum mass (kDa)	Description
Electrospray ionization (ESI)	Nucleic acids, proteins, nonvolatile	HPLC effluent, liquid	200	Soft ionization method
Matrix-assisted laser desorption (MALDI)	Nucleic acids, proteins, nonvolatile	Sample mixed in solid matrix	500	Soft ionization method; very high mass
Fast atom bombardment (FAB)	Carbohydrates, peptides, nonvolatile	Sample mixed in viscous matrix	6	Medium soft ionization method
Electron impact (EI)	Small molecules, volatile	Gas or liquid	1	Hard ionization method
Chemical ionization (CI)	Small molecules, volatile	Gas or liquid	1	Soft ionization method

Table A2.3 A limited list of some types of mass analyzers.

Analyzer	Description
Quadrupole	Unit mass resolution; fast scan; low cost; multiple quadrupoles in series possible
Time of flight (TOF)	High-throughput
Sector (magnetic and/or electrostatic)	High-resolution; exact mass
Ion cyclotron resonance (ICR)	Very high-resolution; exact mass

mass that can be recognized by the ion detector located at the end of the rods, to a point where up to 1000 atomic mass units can be scanned every second. This type of mass spectrometer can also be modified so that a series of quadrupoles can be used, thus increasing the mass resolution.

Several reports have been made on separating small single-stranded RNA and their detection with MS. For example, Bleicher and Bayer [5] used ESI-MS to detect up to 24-mer oligonucleotides separated on a 100 mm-long × 2 mm i.d. Nucleosil C18 column, using a gradient of acetonitrile as solvent with 10 mM ammonium acetate ion-pairing reagent. Bothner *et al.* [6] separated oligonucleotides using an eluent containing 200 mM diisopropylammonium acetate in acetonitrile on a 150 mm-long × 5 mm i.d. column packed with a polymer PLRP-S stationary phase. Phosphodiester, methylphosphate, and phosphorothioate oligonucleotides up to

20-mer in length were each characterized at the 7 nmol level. When Apffel *et al.* [7, 8] utilized a 250 mm-long × 2.1 mm i.d. YMC C18 column to analyze 75-mer oligonucleotides, they found that 100 mM triethylammonium acetate ion-pairing reagent resulted in a drastic reduction in ion formation, and hence a poor mass sensitivity. However, the performance was improved by using a sheath of organic solvent around the eluent fluid so as to improve the stability of the electrospray and enhance the signal intensity of the ESI-MS.

The detection of nucleic acids with ESI may be difficult due to their tendency to form stable adducts with cations such as sodium, the result being mass spectra of poor quality. The removal of sodium cations is necessary in order to obtain high-quality mass spectra; the counterion of the ion-pairing reagent may also drastically affect the result. For example, a multivalent counterion will cause the fragment to elute faster from the column. A number of excellent studies have been conducted by Huber and coworkers [9, 10], where the conditions for separation and detection have been optimized.

A2.3
Data Analysis

The results of the chromatographic separation are generally displayed on a computer screen. The computer uses an A/D (analog to digital) board to convert the analog signal coming from the detector into digital data. The digital information is stored and manipulated to report results to the user. There are three types of information that can be gained by the peaks of the chromatogram. They are the peak retention time, the peak shape or pattern, and the peak size (area and height).

A2.4
Size Analysis

When working with double-stranded RNA, reverse-phase columns and ion-pairing reagents such as triethylammonium acetate (TEAA) will produce size-based separations of the fragments. The retention times from one column to the next, and from one day to the next, are quite reproducible, provided that the column is in good condition and the eluent has been prepared correctly and is not old. The working status of a system can be determined by injecting a pUC18 *Hae*III digest (available from Transgenomic, San Jose, CA and Omaha, NE, USA) and comparing the chromatogram to the separation obtained with the standard. In this way, it is possible to perform what is termed a "zero-point calibration" of the system to determine the sizes of any unknown fragments. WAVE software (Transgenomic) includes default retention time values in the equations used to calculate the size of an unknown fragment, such that it is possible, when using well maintained systems, to accurately calculate fragment size within 10%. A greater degree of accuracy can be achieved by using a 1-point calibration with a single known size standard (preferably multipoint) calibration.

The separations of single-stranded RNA do not proceed only according to fragment size; rather, the fragment sequence also contributes to the retention time. In general, early-eluting peaks will correspond to shorter (degraded) fragments, and in these cases it would most likely not be necessary to run a standard unless there was some reason to confirm the identity of the major peak.

A2.5
Quantification

There are many cases where it desirable to know exactly how much of a particular fragment of RNA is present. For example, downstream processing such as PCR, cloning, and reverse transcription can be performed much more efficiently if the amount of nucleic acid present is known. The absorptivity of double-stranded RNA is about 80% that of single-stranded RNA (i.e., for equal mass amounts of double-stranded and single-stranded RNAs, the absorbance or detector signal will be about 80% lower for double-stranded RNA).

For a given instrument configuration, it is possible that a statement could be made regarding the area/mole ratio for a typical analyte. However, much better results could be obtained if a standard were to be run at least once on the configuration used to determine the area/mole relationship. If the instrument is functioning properly, it is possible to calculate concentration at least to within 50% with no (i.e., zero-point) instrument calibration. More accurate investigations require the use of a standard of known concentration.

Recall that the whole molecule is detected with UV detection, and that either the molar or mass concentration or mass amount can be calculated. With fluorescent detection, molar concentration is usually calculated. (The mass concentration can be calculated using the molecular weight of the particular fragment being measured.) By using a simple ratio, the calculation of the concentration can be performed as:

$$\text{unkonwn conc.}/\text{unkown peak area} = \text{known conc.}/\text{known peak area} \quad (A2.2)$$

and therefore,

$$\text{unkonwn conc.} = \text{konwn conc.}/\text{known peak area} \times \text{unknown peak area} \quad (A2.3)$$

It is also possible to draw a calibration curve of known peak area versus known concentration, thereby finding the unknown peak area on the curve and measuring the unknown concentration. The majority of this manipulation is performed automatically by software, once a method has been installed. Although the multipoint calibration curve is the most accurate method, this level of accuracy is rarely needed and a single point calibration ratio is usually sufficient.

In order for the data system to measure peak area or peak height, the baseline of the peak must be accurately drawn. Whilst the software will attempt to draw a baseline for the peak, the user must also frequently manually mark the baseline

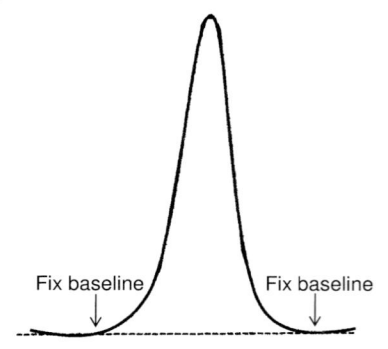

Figure A2.12 The baseline is marked on each side of the peak, as shown in the figure. Fixing of the peak baseline may have to be performed manually with a mouse using the peak area measurement software.

start and finish points so as to draw the peak baseline accurately. It is a common mistake of new users to trust the software to draw the correct baseline. The correct way to draw baselines for peak integration (measurement of peak area) is shown in Figure A2.12.

It is often necessary to convert moles of material concentration to mass of material. For double-stranded RNA, the conversion of pmol to µg is as follows:

$$\text{pmol} \times N \times 660 \left(\text{pg mol}^{-1} \text{ base pairs}\right) \div 1 \times 10^{-6} \text{ pg } \mu g^{-1} = \mu g \qquad (A2.4)$$

where N is the number of base pairs.

Sample preparation and injection volume errors probably account for the majority of the short-term variation in area measurements. Poor pipette techniques may also create some errors.

A2.6
Fragment Collection

One of the most powerful features of RNA Chromatography is that the material can easily be purified by collecting directly from the detector effluent. Although such collection can be performed by hand, it is invariably accomplished using an automated fragment collector and controlling software, with the material being collected into either single vials or large PCR plates.

Care should be taken to ensure that the actual peak has been collected. Measurements of recovery are carried out by taking a small portion of the recovered peak, re-injecting and measuring the area, multiplying this by the ratio of total collected volume to re-injected volume, and then comparing this value to the area of the original peak. Normal recoveries are approximately 80%. Of course, materials such as RNA may be degraded (enzymatically) after they have been collected; likewise, high recoveries may not be possible due to loss of the material during the concentration process (by precipitation or solvent evaporation).

One very important parameter in fragment collection is the careful execution of the *collection time*, with problems generally arising as a result of the lag period that occurs between the time that the peak is detected and when the fragment is deposited in the collector. The most reliable collection method is "timed collection", although if the timing is incorrect it is easily possible to miss some – or even all – of the peak. It is also important for there to be as little dead volume as possible in the tubing between the detector cell outlet and the tip of the deposition probe. A too-large dead volume can destroy the resolution of the separation and also result in cross-contamination of the peak of interest with neighboring peaks [11]. It is also very important that the probe is cleaned between the collection of each peak; this is normally carried out automatically by the fragment collector.

References

1 Sutton, J.E., Gjerde, D.T. and Taylor, P.D. (2000) MIPC chromatographic apparatus with improved temperature control, U.S. Patent 6,103,112.

2 Oefner, P.J., Huber, C.G., Umlauft, F., Berti, G.N., Stimpfl, E. and Bonn, G.K. (1994) High-resolution liquid chromatography of fluorescent dye-labeled nucleic acids. *Anal. Biochem.*, **223**, 39.

3 Becker, K.H., Taylor, P.D. and Gjerde, D.T. (1999) Mutation detection by denaturing DNA chromatography using fluorescently labeled polymerase chain reaction products. *Anal. Biochem.*, **272**, 156.

4 Willard, H., Merritt, L., Dean, J. and Settle, F. (1998) *Instrumental Methods of Analysis*, 7th edn, Wadsworth Publishers, Stamford, CT.

5 Bleicher, K. and Bayer, E. (1994) Analysis of oligonucleotides using coupled high performance liquid chromatography–electrospray mass spectrometry. *Chromatographia*, **39**, 405.

6 Bothner, B., Chatman, K., Sarkisian, M. and Siuzdak, G. (1995) Liquid-chromatography mass-spectrometry of antisense oligonucleotides. *Bioorg. Med. Chem. Lett.*, **5**, 2863.

7 Apffel, A., Chakel, J.A., Fischer, S., Lichtenwalter, K. and Hancock, W.S. (1997) Analysis of oligonucleotides by HPLC–electrospray-ionization mass spectrometry. *Anal. Chem.*, **69**, 1320.

8 Apffel, A., Chakel, J.A., Fischer, S., Lichtenwalter, K. and Hancock, W.S. (1997) New procedure for the use of high-performance liquid chromatography–electrospray mass spectrometry for the analysis of nucleotides and oligo-nucleotides. *J. Chromatogr.*, **777**, 3.

9 Huber, C.G. and Krajete, A. (1999) Analysis of nucleic acids by capillary ion-pair reversed-phase HPLC coupled to negative-ion electrospray ionization mass spectrometry. *Anal. Chem.*, **17**, 3730.

10 Premstaller, A., Oberacher, H. and Huber, C.G. (2000) High-performance liquid chromatography-electrospray ionization mass spectrometry of single and double-stranded nucleic acids using monolithic capillary columns. *Anal. Chem.*, **72**, 4386.

11 Hecker, K.H. and Kobayashi, K. Application Note 115. Sequencing of DNA fragments isolated with the WAVE® nucleic acid fragment analysis system. Available from Transgenomic Inc., Omaha, NE.

Appendix 3
RNA Chromatographic System Cleaning and Passivation Treatment

A3.1
Background Information

HPLC separation problems caused by corroding stainless steel surfaces have been reported in the analysis of inorganic ions and proteins. Poor electrochemical detection (i.e., detection by oxidation or reduction of the sample peak) has sometimes been shown to be due to metal ion contamination found on, or released from, stainless steel surfaces [1]. Most HPLC chromatographic systems are constructed from metal components, and RNA Chromatography has extremely high requirements for system cleanliness in order to achieve the reliable separation and detection of nucleic acid fragments. It is essential that the surfaces of the system do not contain metal ions that will either trap the nucleic acids or cause them to undergo structural or conformational changes. Neither must the surfaces release metal ions that can travel to others point within the system (e.g., the column end inlet frit), and in turn harm the separation [2–5].

Fortunately, it is possible to clean and maintain stainless steel components and other (plastic or metal) materials that contain harmful metal ions [6, 7] and, indeed, most manufacturers of HPLC equipment provide standard operating procedures on how to clean and passivate their equipment. These original procedures have been further modified to ensure that the RNA Chromatography system surfaces are contamination-free and stable. In general, treatment with nitric acid will clean and passivate a stainless steel surface, rendering it resistant against further corrosion. Although the concentration of nitric acid used for HPLC system passivation may vary from $3\,M$ to $13\,M$, the use of $8\,M$ acid is generally sufficient for the rapid and effective cleaning of the RNA Chromatography system.

During the cleaning process, the nitric acid will remove existing metal ion contamination as well as some organic contaminants. Typically, nitric acid is used to clean organic and metal materials on quartz detector windows in UV detectors, and also to remove surface corrosion from the system's tubing. However, the nitric acid procedure should not be used to clean incompatible components. For example, a (rust-prone) stainless steel frit is not a suitable component for a RNA chromatograph. Moreover, a system that has worked well for protein separations may not necessarily work for RNA separations. Stainless steel is an excellent material for

RNA Purification and Analysis: Sample Preparation, Extraction, Chromatography
Douglas T. Gjerde, Lee Hoang, and David Hornby
Copyright © 2009 WILEY-VCH Verlag GmbH & Co. KGaA, Weinheim
ISBN: 978-3-527-32116-2

HPLC systems, and is used where high structural strength and reliability are required. However, stainless steel frits (which have a very large surface area and are prone to corrosion) should be replaced with frits made from titanium or polymers, rather than relying on cleaning the system to maintain its performance. Even then, titanium or polymers may contain metal ion contamination that must be removed.

During the passivation process, the nitric acid produces a very thin oxide layer on the metal surface that will protect the metal from further attack by oxygen, acid, and other influences. Nitric acid will also preferentially remove iron from the surface (and also from any sharp edges that are prone to corrosion), effectively leaving chromium and nickel metal at relatively higher concentrations at the surface. Fortunately, both of these metals are less likely than iron to undergo further corrosion.

In the procedure described below, treatment with a chelating reagent (ethylenediamine tetra-acetic acid; EDTA) is included to ensure complete cleaning and passivation. Such a system treatment should be performed once every 6 months, or whenever a deteriorating column performance (broad, inconsistent or missing peaks) is encountered. It is also recommended that the piston seals, injection port seals and flow path filter are all changed after the passivation process.

If a HPLC system becomes contaminated, then even a new column can show a deterioration in performance. In fact, if the performance of a new column is shown to degrade very quickly, it is a sure sign that the system requires cleaning.

> **A note of caution**: Nitric acid can be dangerous. ALWAYS wear safety glasses and gloves. The nitric acid effluent should be collected in a separate (empty) waste container. Do NOT mix concentrated nitric acid with organic solvents of any kind.

The person performing the passivation must be familiar with the basic operation of the RNA Chromatography system, and must know how to remove and install the injection port and seal, and the injection needle. Purging and rinsing may have to be carried out in manual mode with 50% A and 50% B, so as to passivate both reservoir flow paths. The column and inline degassers should be removed from the flow path to prevent them from being harmed. The detector should also be removed, unless there is a need for this component to be cleaned. The recommended flow rate for cleaning and rinsing is $2\,\text{ml}\,\text{min}^{-1}$. The door to the column heater should NOT be closed; neither should the heater be turned off. All passivation procedures should be performed at room temperature.

In an experiment performed to demonstrate the effect of passivation on a new system that had not yet undergone final cleaning and conditioning, identical columns were used for all four chromatograms. Figures A3.1 and A3.2 show how a contaminated system might perform for a size standard and a mutation standard. In these cases, the peak pattern is unusual and the resolution of some peaks is poor. In other, more severe, cases the peaks may split into doublets or even disappear. Figures A3.3 and A3.4 show how the separation should look in a clean system. In

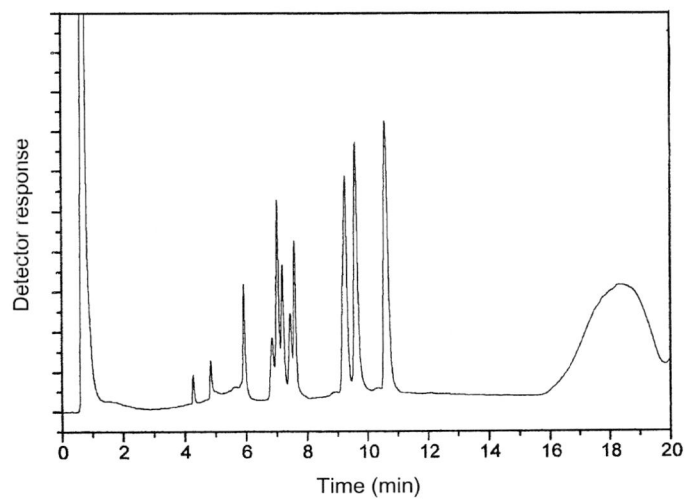

Figure A3.1 Separation of pUC 18 HaeIII digest size standard on a new DNA Chromatography system before system treatment. The separation conditions are specified as standard test conditions for performing a size-based separation for a new DNASep column. (From Transgenomic, Inc., with permission).

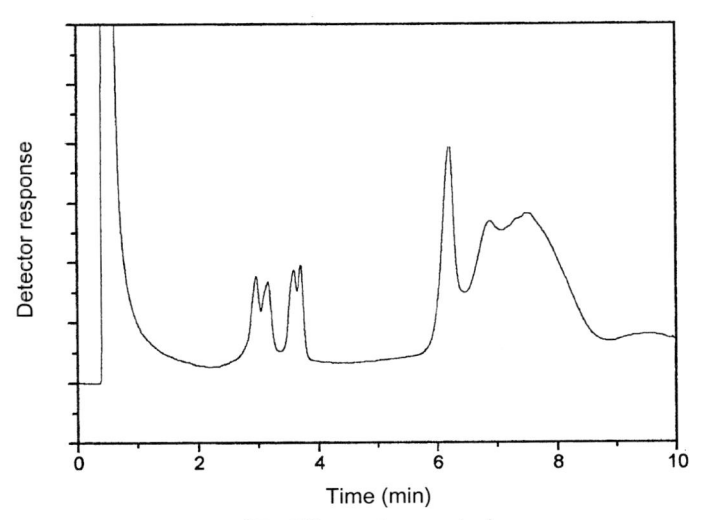

Figure A3.2 Separation of Dys271 mutation standard on new DNA Chromatography system before system treatment. The separation conditions are specified as standard test conditions for performing a size-based separation for a new DNASep column. (From Transgenomic, Inc., with permission).

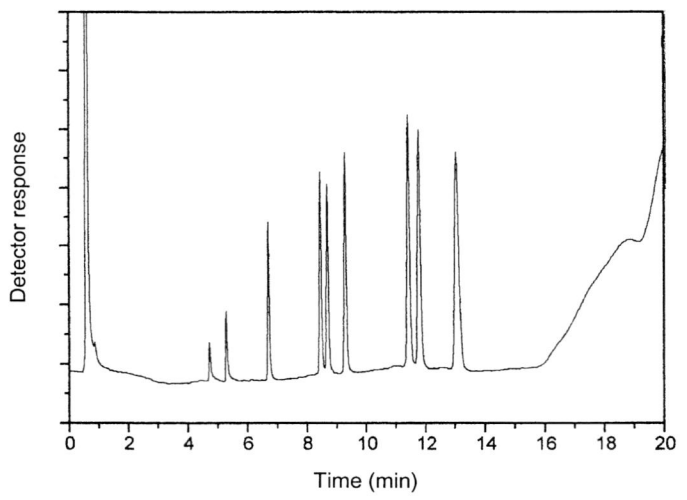

Figure A3.3 Separation of pUC 18 *Hae*III digest size standard on a new DNA Chromatography system after system treatment. The separation conditions are specified as standard test conditions for performing a size-based separation for a new DNASep column. (From Transgenomic, Inc., with permission).

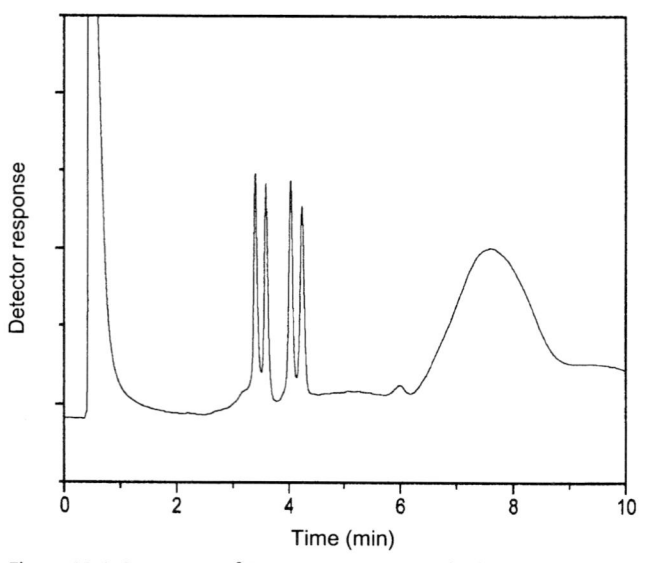

Figure A3.4 Separation of Dys271 mutation standard on new DNA Chromatography system after system treatment. The separation conditions are specified as standard test conditions for performing a size-based separation for a new DNASep column. (From Transgenomic, Inc., with permission).

these cases, the peak patterns are normal (as compared to the standard chromatograms included with the new column), and the peak resolution is excellent. This cleaning and passivation procedure should not to be considered as an exotic rescue attempt, but rather as a simple way of preventing premature system failure.

A3.2
Reagents

- Deionized (DI) water
- 8 M nitric acid (ca. 35%, w/w)

    Preparation: To a 250 ml media bottle, add slowly 70 ml of concentrated nitric acid (HNO_3; 69.5%, $d = 1.4 \text{ g ml}^{-1}$) to 100 ml of DI water. The solution will become warm; wait until the acid has reached room temperature before using.

- EDTA (tetrasodium salt), 50 mM

    Preparation: Dissolve 22.6 g of tetrasodium EDTA (MW = 452.24) in 500–700 ml of DI water (use a 1 l volumetric flask). When all the solids have dissolved, top up with DI water to the 1 l mark.

A3.3
Preparation of the System

- Replace the column with a PEEK union.
- Bypass the detector flow cell with Teflon tubing (0.010 or 0.020″ i.d.) from the preheater coil to a new waste container of appropriate length.
- Rinse the entire system with distilled water (all channel reservoirs): purge for 5 min and flush for another 30 min at 2 ml min^{-1}. Check for any leaks! Make sure that the system is without leaks. Do NOT close the door of the column oven, or turn off the oven.

A3.4
Passivation of the System

- Remove the white solvent inlet filter caps from the solvent lines of all channel reservoirs.
- Fill 8 M nitric acid into an appropriate glass bottle (Erlenmeyer flask, media bottle) and insert the tubing ends of all channel reservoirs into the acid.
- Rinse all channel reservoirs with 8 M nitric acid: purge for 5 min; then rinse the system for another 15 min at 2 ml min^{-1}.

- Replace the nitric acid with DI water. Rinse all channel reservoirs with DI water: purge for 10 min and flush for 90 min at 2 ml min^{-1}. During this time, change the DI water at least three times.

- Check the pH of the effluent with pH paper. Keep rinsing with water until the pH is >5.

- Replace the DI water with a 50 mM solution of tetrasodium EDTA. Purge for 5 min and rinse the system for 30 min at 2 ml min^{-1}. **Caution**: If the nitric acid is not rinsed away thoroughly, the EDTA might precipitate and plug the system.

A3.5
Equilibration of the System

- Install new inlet filter caps.

- Replace the EDTA solution with water, purge for 5 min, and rinse for 15 min.

- Replace the water with fresh eluents (A and B), purge for 5 min, and rinse for at least 2 h at 2 ml min^{-1}.

- Install the column and equilibrate it with at least one gradient run. Test the system with a pUC 18 *Hae*III digest and a Dys271 mutation standard.

References

1 Collins, K.E., Collins, C.H. and Bertran, C.A. (2000) Stainless steel surfaces in LC systems, Part I – corrosion and erosion. *LC-GC*, **18**, 600.
2 Gjerde, D.T., Haefele, R.M. and Togami, D.W. (2000) Method for performing polynucleotide separations using liquid chromatography, U.S. Patent 6,017,457.
3 Gjerde, D.T., Haefele, R.M. and Togami, D.W. (1998) System and method for performing polynucleotide separations using liquid chromatography, U.S. Patent 5,772,889.
4 Gjerde, D.T., Haefele, R.M. and Togami, D.W. (2000) Apparatus for performing polynucleotide separations using liquid chromatography, U.S. Patent 6,030,527.
5 Gjerde, D.T., Haefele, R.M. and Togami, D.W. (1999) Liquid chromatography systems for performing polynucleotide separations, U.S. Patent 5,997,742.
6 Collins, K.E., Collins, C.H. and Bertran, C.A. (2000) Stainless steel surfaces in LC systems, Part II – passivation and practical recommendations. *LC-GC*, **18**, 688.
7 Shoup, R. and Bogdan, M. (1989) Passivation of liquid chromatography components. *LC-GC*, **7**, 742.

Index

a
A term 155
absorbance unit (AU) 174
adsorption of sample compounds and sample matrix compounds 37
affinity chromatography 53
agarose matrix 53
alkylammonium salt 10
alternative splicing 120
amino-acyl tRNA (aa-tRNA) 73f., 131
aminoacyl tRNA synthetase (ARS) 132
anion exchanger 52, 154f.
anion-exchange resin 50
– silica-based 52
aptamer 33
Arabidopsis 31
autosampler injector 168

b
B term 155
bacteria, RNA 24
biological function 17
– genetic approach 17
– systems biology approach 17
biological process
– analysis 3
– chemical reaction 17

c
C term 156
Caenorhabditis elegans 30
capacity factor 147ff.
carrier precipitate 38
cartridge, *see* separation column
catalytic RNA 22
cDNA library synthesis, DNA and RNA chromatography 114
cell culture 31
cellulose affinity substrate 53

central dogma 127ff.
cetyltrimethylammonium bromide (CTAB) 31
chaotropic denaturing interaction mechanism 61
chemical probing 140ff.
chromatographic separation
– data analysis 180
– equation 147ff.
– size analysis 180
chromatographic term 148
chromatography 4, 147ff.
– plate theory 151
– RNA 81ff., 101ff.
classification, RNA 17ff.
cleavage 140
collection time 183
column
– extra-column effect 157
– guard 170
– oven 85, 171
– protection 170
– scavenger 170
– type 6
competitive RT-PCR product, DNA chromatography 117
complementary DNA (cDNA) 135
– library 114
covariation 137
crosslink 44

d
dead time 148
dead volume 161
denaturing HPLC (DHPLC) 89
density gradient 71
DEPC (diethyl pyrocarbonate) 87
depletion column 38

detection
– limit   172
– RNA chromatograph   172
detector sensitivity   172
dextran   53
dicer   76
diethylaminoethyl anion exchanger (DEAE)   48
difference-detecting engine   89
differential display (DD)   121
differential messenger RNA display   121ff.
– DNA chromatography   121
directed hydroxyl radical probing   143
DNA
– size-based separation of double-stranded DNA   107
– transcription   67
DNA chromatography   89, 114ff.
– differential messenger RNA display   121ff.
– RT-PCR product   117
DNA footprinting   139
Drosophila melanogaster   31

e

eddy diffusion   155
eluent
– chemistry   6
– degassing   162
elution
– gradient   57f.
– isocratic   57ff.
– shallow gradient   57f.
– volume   64
Escherichia coli   23
– RNA   23
eukaryote   28
eukaryotic gene expression, post-transcriptional control   76
extra-column effect   157

f

fluorescence   174
– detector   174
fluorescent detection   86
fluorescent dye   174f.
fluorescent tag   86, 176
footprinting   139ff.
N-formyl-methionyl-tRNA^{fMet} (f-Met-tRNA^{fMet})   73
Förster resonance energy transfer (FRET)   139ff.
fragment collection   86, 182
full loop injection   168

functional group   39f.
– attachment   43
fungi   30

g

gel electrophoresis   90
– agarose   92
– comparison of RNA chromatography   94
– polyacrylamide   92
gel filtration   63
gel filtration chromatography   53, 63
gel filtration material   54
– separation size range   54
gel permeation   53
gene function   79
gene regulation   76
– microRNA (miRNA)   78
general detector   172
gradient
– blank   163
– continuous   6
– formation   164
– shallow   57f., 88
– step   6
– type   56
guanidinium chloride   61f.
guide RNA (gRNA)   27

h

height equivalent of a theoretical plate   148ff.
HEMA   49
high-performance liquid chromatography (HPLC)   10f., 147ff.
– cleaning   185
– instrumentation   159ff.
– passivation   186ff.
– RNA chromatography   81
– silica-based   41
high-pressure gradient mixing   166
human telomerase RNA (hTR)   95
– nondenaturing condition   95
hybridization   62
hydroxyl radical probing   139

i

in vitro reconstitution assay   75
in vitro transcription system   68
in vitro transcription assay   68
initiation complex   73
injection valve   168
interaction type   40
internal loop   138
internal ribosome entry site (IRES)   76

ion-exchange chromatography 59, 154
ion-exchange separation mechanism 57ff.
ion-pairing on a reverse-phase substrate 56
ion-pairing, reverse-phase liquid chromatography 82

l

liquid chromatography 82, 93
liquid volume 63
low-pressure gradient mixing 166

m

mass spectrometry detection 177
mass spectrometry ionization method 178f.
messenger RNA (mRNA) 21ff.
– separation 108
microarray hybridization technique 62
microRNA (miRNA) 27, 78, 128
– gene regulation 78
mobile phase gradient 5
model building 139
modification interference 143
monolith polymeric column 47
monomer mix 46
multicellular organism 30
multipath 155

n

non-coding RNA (ncRNA) 27
nucleotide analogue interference mapping (NAIM) 139

o

oligoribonucleotide, synthetic 111

p

partial loop injection 168
peak broadening 157
peak capacity 153
peak diffusion 155
peak resolution 148
peak width 148ff.
peptidyl-tRNA 74
phase
– mobile 4ff., 151ff.
– stationary 4f., 151ff.
plant 31
plate theory, chromatography 151
polyacrylamide gel electrophoresis (PAGE) 14, 92
polyacrylamide gel filtration substrate 53
polyacrylate polymer 49
polymer, functionalization 48
polymeric resin substrate 44
– porous and nonporous 45ff.
precursor messenger RNA (pre-mRNA) 25, 128
– splicing 71
pressure, RNA chromatograph 167
primary mature RNA (pri-mRNA) 25, 72
primary transcript RNA (pri-miRNA) 27
prokaryote 28
pseudoknot 138
pump, RNA chromatograph 159ff.

q

quantification, RNA chromatography 181

r

resin capacity 154
retention 147f.
retention factor 148ff.
reverse phase ion-pairing separation mechanism 54
reverse phase liquid chromatography 82, 154
ribonuclease enzyme 14
ribosomal RNA (rRNA) 21
– 16S rRNA 75
– separation 108
30S ribosomal subunit 133
50S ribosomal subunit 133
– reconstitution 75
ribosome 73
70S ribosome initiator complex 73
ribozyme 13, 22, 69, 133
– hairpin 113
RISC, *see* RNA-induced silencing complex
RNA 12
– analysis 3, 67ff.
– biological classification 19
– chemical classification 18
– classification 17ff.
– enzymatic treatment 13
– eukaryotic cellular 24ff.
– fluorescently labeled 112f.
– poly(A) tail 62
– prokaryotic cellular 20ff.
– type 19ff.
RNA affinity chromatography 7
RNA catalysis 69
RNA chromatograph 159ff.
RNA chromatography 6ff., 81ff.
– analysis 109ff.

- cDNA synthesis   114ff.
- cleaning   185
- comparison of gel electrophoresis   94
- condition   87
- double stranded   88, 102ff., 180f.
- feature   101
- instrumentation   85, 159ff.
- liquid chromatography   93
- passivation treatment   186ff.
- separation   101ff.
- single-stranded   88, 102ff., 181
- temperature   171
- temperature mode   88
- total RNA extract   107
RNA editing   27
RNA extraction   1ff., 56, 67ff., 79
RNA footprinting   112, 139ff.
RNA high-performance liquid
  chromatography (HPLC)   10f., 41,
  81ff.
RNA interference (RNAi)   26
- pathway   76
RNA purification   71
- commercially available kit   78
- density gradient   71
RNA separation   1ff., 37ff., 82
- cellular RNA   107
- double-stranded   88, 102ff.
- fully denaturing   90
- messenger RNA   108
- nondenaturing   89
- partially denaturing   89
- principle   147ff.
- ribosomal RNA   108
- separation column   169f.
- single-stranded   88, 102ff.
- total RNA   108
RNA structure   14, 41, 130
- determination   135
- primary   136
- quaternary   138
- secondary   136
- tertiary   137
RNA structure–function probing   127
RNA-induced silencing complex (RISC)
  26, 77, 128
RNA–protein complex interaction   69
RNase   61
- chemical probing   142
rock, RNA   32
RT-PCR product, DNA chromatography
  117
RUSH, 1α and 1β isoform   121

S
selective detector   172
SELEX (systematic evolution of ligands
  by exponential enrichment)   34, 62
separation, *see also* RNA separation   152
separation column   159, 169f.
separation factor   148
Shine–Delgarno sequence   73
short hairpin RNA   34
short interfering RNA/small interfering
  RNA (siRNA)   26, 76ff.
signal recognition particle (SRP)
  complex   22
signal recognition particle RNA
  (srpRNA)   22
silica anion exchanger   52
silica hydrogel bead   50
silica material
- functionalization   51
- pellicular   51f.
- porous and nonporous   51f.
silica–glass-based substrate   50
size analysis, RNA chromatography   180
size exclusion   63
small hairpin RNA (shRNA)   27
small interfering RNA (siRNA), *see* short
  interfering RNA
small nuclear ribonucleic protein (snRNP)
  26, 72
small nuclear RNA (snRNA)   26, 72
small nucleolar ribonucleoprotein
  (snoRNP)   25
small nucleolar RNA (snoRNA)   25
soft animal   30
soil, RNA   32
solid-phase extraction   8
solid-phase interaction   37ff.
solid-phase substrate   39ff.
solid surface interaction   41
spin-column   9
spliceosome   26
splicing, alternative   120
stem loop   138
step-gradient process   6
structure–function paradigm   127
substrate   40
sucrose gradient   70
synthetic RNA   32

t

TEAA (triethyl ammonium acetate)   11, 55, 83
telomerase RNA   24, 105
– human (hTR)   95
Tetrahymena, nuclear extract   69
tetraloop bulge   138
tetramethylammonium chloride (TMAC)   59
theoretical plate number   148
total RNA, separation   108
transcription   67
transfer-messenger RNA (tmRNA)   22
transfer RNA (tRNA)   20, 131
– analysis   109
translation   72

u

ultraviolet (UV) detection   86
ultraviolet-visible (UV-VIS) detector   173

v

van Deemter equation   155
van Deemter plot   156
vault RNA (vRNA)   22
virus   32
– RNA   76
volume
– dead   161
– elution   64
– interstitial or void   64
– liquid   64
vRNA–protein complex (vRNP)   22

w

Watson–Crick complementarity   137
wobble rule   139

y

yeast   29